Management Principles and Practices for Technical Communicators

THE ALLYN AND BACON SERIES IN TECHNICAL COMMUNICATION

Series Editor: Sam Dragga, Texas Tech University

Management Principles
and Practices
for Technical Communicators

R. Stanley Dicks

North Carolina State University

New York   San Francisco   Boston
London   Toronto   Sydney   Tokyo   Singapore   Madrid
Mexico City   Munich   Paris   Cape Town   Hong Kong   Montreal

Vice President: Eben W. Ludlow
Executive Marketing Manager: Ann Stypuloski
Production Manager: Eric Jorgensen
Project Coordination, Text Design, and Electronic Page Makeup: Electronic Publishing Services Inc., NYC
Cover Designer/Manager: John Callahan
Manufacturing Buyer: Roy Pickering
Printer and Binder: Hamilton Printing Co.
Cover Printer: The Lehigh Press, Inc.

Library of Congress Cataloging-in-Publication Data
Dicks, R. Stanley.
    Management principles and practices for technical communicators / R. Stanley Dicks.--1st ed.
        p.   cm.
    Includes bibliographical references and index.
    ISBN 0-321-16523-3 (pbk.)
    1. Communication of technical information--Management. I. Title.

T10.5.D53 2003
808'.066--dc21

                                                2003043871

Please visit our Web site at **http://www.ablongman.com/**

ISBN 0-321-16523-3

1 2 3 4 5 6 7 8 9 10—HT—06 05 04 03

CONTENTS

Management Philosophy Issues Related to Technical Communication   1

CHAPTER

3 Finance for Technical Communication Managers   111

CHAPTER

4 Management Communication   153

Management Training   186

CHAPTER 6

Managing Yourself   207

CHAPTER 7

Managing Your Boss   220

CHAPTER

8

Other Technical Communication Management Responsibilities   230

CHAPTER

9

Resources for Technical Communication Managers   246

FOREWORD
by the Series Editor

The Allyn and Bacon Series in Technical Communication is designed to meet the continuing education needs of professional technical communicators, both those who desire to upgrade or update their own communication abilities as well as those who train or supervise writers, editors, and artists within their organization. This series also serves the growing number of students enrolled in undergraduate and graduate programs in technical communication. Such programs offer a wide variety of courses beyond the introductory technical writing course–advanced courses for which fully satisfactory and appropriately focused textbooks have often been impossible to locate.

The chief characteristic of the books in this series is their consistent effort to integrate theory and practice. The books offer both research-based and experienced-based instructions, describing not only what to do and how to do it but explaining why. The instructors who teach advanced courses and the students who enroll in these courses are looking for more than rigid rules and ad hoc guidelines. They want books that demonstrate theoretical sophistication and a solid foundation in the research of the field as well as pragmatic advice and perceptive applications. Instructors and students will also find these books filled with activities and assignments adaptable to the classroom and to the self-guided learning processes of professional technical communicators.

To operate effectively in the field of technical communication, today's technical communicators require extensive training in the creation, analysis, and design of information for both domestic and international audiences, for both paper and electronic environments. The books in the Allyn and Bacon Series address those subjects that are most frequently taught at the undergraduate and graduate levels as a direct response to both the educational needs of students and the practical demands of business and industry. Additional books will be developed for the series in order to satisfy or anticipate changes in writing technologies, academic curricula, and the profession of technical communication.

Sam Dragga
Texas Tech University

PREFACE

This book is based on fifteen years of experience managing groups of technical communicators in various environments and on a thorough study of the books, articles, and presentations covering management topics in the technical communication literature.

All technical communicators have to practice project management to at least some extent. Many project teams have a lead writer who is required to exercise project management responsibilities but who may not have supervisory responsibilities. In many organizations, in fact, there are no technical communicators in supervisory positions because they report to managers in other organizations, often development, engineering, or marketing. Most of the literature written about management for technical communicators, therefore, focuses more on project management than it does on other aspects of managing an organization, such as management philosophy, organizational structure, managing personnel, managing finances and budgets, management communication, career development, etc. Many articles have been written about technical communication project management, and JoAnn Hackos's excellent 1994 book, *Managing Your Documentation Projects*, covers the subject thoroughly.

This book treats the other areas in a systematic way, suggesting principles for consideration in each area, followed by analyses of the practices that work for managers and identification of those practices to avoid. This book does not attempt to cover all aspects of management, as the "MBA handbook"-style texts do. Rather, it focuses on those aspects of managing a technical communication group that differ from what one can find in general management texts.

Intended Audience

The audience for this book comprises a considerable diversity of technical communicators. Obviously, the primary audience includes currently practicing managers of technical communication functions. Because the discipline is relatively young and has grown so fast, many practitioners have been put into management positions in the last few years. Many of them have had no formal education, and often, precious little training in how to carry out basic management functions or to conceptualize their roles. While the major emphasis of this book is to address the needs of this audience, several other groups are also addressed, including:

- Managers from other disciplines who find themselves responsible for a technical communication function

- Employees who are not in management positions but who contemplate being promoted into such a position eventually
- Entrepreneurs who are managing their own technical communication consulting companies
- Students, usually in graduate programs, who are taking a class or a segment of a class aimed at teaching them about technical communication management
- Professors who teach such classes.

The Central Premises of the Book

This book has four main premises:

Keep eye on big picture

1. Technical communication managers need to conceptualize what they do, both in a larger, overall sense related to their organizational roles and also in a more detailed, day-to-day sense related to each of the tasks they perform. While lists of management skills are helpful, it is essential to consider each management role and responsibility in its larger organizational, social, political, inter-personal, and rhetorical environments.

Work does not exist in a vacuum

2. Technical communication managers must understand and participate in overall organizational missions and goals, and must align their organizations and operations to participate in achieving those goals.

3. Technical communicators are product and service developers who must participate fully and equally with other developers (Shirk 1989; Johnson-Eilola 1996).

Advocate value to company "PR" person

4. Technical communication managers must educate peers and upper management about the value that their groups add to the development process. They must constantly advocate better working processes (Hackos 1989, 1994) and reward systems for their disciplines.

The book also proceeds from two significant assumptions: (1) that achieving consistent, high-quality technical communication performance is possible only with carefully managed, highly structured quality processes using established and proven management methods, and (2) that in the information age, many traditional principles and methodologies are being made obsolete. New principles and practices are necessary for managing employees in the information economy (Gatien 1990), and many of those new practices (managing remotely, for example) are under constant revision and improvement. This requires that managers undertake continual education and training to keep abreast of multiple technologies and their effects on technical communication functions.

The Structure of the Book and How to Use It

The order of the sections is somewhat arbitrary. However, I have tried to structure them roughly in the order that would most interest a new technical communication manager. This means that management philosophy comes first, closely followed by personnel management, often the most difficult task for new managers. Following are sections on financial management, communication, career development, managing oneself, and special situations.

I have tried to organize the book so that it works when read straight through as a linear text, but also so that it is effective as a reference work for a manager who is looking for pointers or reminders related to a specific task. To that end, the sections are broken down into numerous sub-sections, each with a heading describing a particular idea, task, or sub-task.

Beginning managers and students will benefit from reading the book straight through. More experienced managers might better skip around, reading the subjects that interest them the most or that offer some extra information. To the extent possible, the sections are self contained, with few cross references to other sections. That makes it possible for experienced managers to read sections in isolation and get as much out of them as possible as quickly as possible. Checklists and forms throughout the book are intended for use while conceptualizing, planning, and performing various tasks.

The references used in this book are primarily from the technical communication literature. There is, of course, a huge body of literature about management, but I have tried to limit the citations, whenever possible, to books and articles specifically devoted to technical communication management issues.

At the end of each chapter there is a set of discussion questions, followed by a case-based writing assignment.

Acknowledgements

One learns about management from many sources and role models. Each of the managers we work with teaches us something, sometimes by positive example and sometimes otherwise. I would like to thank all of the supervisors to whom I've reported, especially Bob Hughey, whose positive influences inform many sections of this book.

My graduate students in our managing technical communication course have consistently challenged me to rethink management ideas and to describe concepts and practices in ways that help them conceptualize and carry out their jobs. I also thank the alumni who have contacted me and told me what worked and what did not, and what new concerns they had once they became managers.

My colleagues in the Rhetoric and Technical Communication program at North Carolina State have provided valuable assistance in helping me track down sources, however obscure. They have also given me excellent advice about finishing the book and possible venues for its publication. Special thanks to Brad Mehlenbacher and Susan Katz who have always been willing to discuss the management ideas I presented to them.

Sam Dragga, the series editor, offered valuable early advice and encouragement, as did Eben Ludlow, Vice President at Pearson Longman. Lisa Kinne's firm efficiency made the production phase a delight rather than a burden.

The reviewers listed below to whom Longman sent the book offered excellent comments and suggestions for improvements and additions, many of which have been incorporated.

Stephen A. Bernhardt, University of Delaware
Meredith Blackwelder
Brenton Faber, Clarkson University
Mark Haselkorn, University of Washington
George F. Hayhoe, Mercer University
Catheryn L. Mason, SolArc, Inc.

I would like to thank my parents, both of whom worked in management positions during their careers. My father, Bob, has been especially responsible for teaching me to think about the kinds of problems that managers face and for showing me at an early age how to think about management in ways that are consistent with one's principles.

Finally, thank you, most of all to Janet, without whose cheerful encouragement and patience this book would not have been possible.

R. Stanley Dicks
North Carolina State University

CHAPTER 1

Management Philosophy Issues Related to Technical Communication

Introduction

The idea of management philosophy may seem alien to those who have been plagued with managers who did not evince any signs of having one. We do not ordinarily associate philosophical thought as being consistent with the ways in which managers process information and act on it. However, it is important for technical communication managers to think philosophically about a number of issues, at the very least those related to ethics and at best those related to their overall function as managers.

This section is not intended to provide comprehensive treatment of management philosophies. Such treatment is impossible in a single chapter, and indeed, many large tomes on the issue have been published. Rather, it will briefly discuss some of the philosophical stances that managers adopt to guide them in doing their jobs and will suggest ways to evaluate those stances and decide which are appropriate for them and their situations.

The new manager needs to pay special attention to the question of management philosophy. It is far too easy to begin managing by reacting to events and people without any underlying set of tenets and ideals. Indeed, many managers do not develop any sense of philosophy until they have been in management for years and have realized the need to do so. Many never do it at all. But the best managers, whether formally or informally, consciously or intuitively, establish a set of principles to guide the way they function. While new managers might not be ready to establish such principles for themselves, they should at least be aware of the need to do so and of some of the methods that will assist them in considering philosophical questions.

This chapter discusses management philosophy by addressing three concerns: (1) principle- and value-based management models and their value for technical communicators, (2) mission statements or vision statements and how they can help a manager and an organization focus on its most important philosophical tenets, and (3) the various organizational structures for technical communication groups and how they reflect management philosophy regarding technical communication within the organization.

Principle- and Value-Centered Management

The idea of principle-centered management has come from Stephen R. Covey through his books, *The 7 Habits of Highly Successful People* and *Principle-centered Leadership*. The basic concept is that a manager leads by following certain principles that extend not only to work life but also to life outside of work. A principle-centered manager knows what his/her personal goals are, both in private life and at work, and has written those goals down, both for the long term and for the short term. In other words, this is not a "contemplate your navel" philosophy, but rather one that requires active effort throughout.

Any time we begin to talk about principles, we rapidly get into the difficult areas of ethics, morality, and religion. Covey largely avoids this by essentially referring to what he calls "natural principles," which are often labeled by others as "natural law." The basis for principle-centered management is that the manager has formulated his/her basic set of guiding principles, used them to inform a set of short-term and long-term goals, and uses the principles and goals as guidance for making management decisions in personal life and in work life. Principle-Centered Leadership is divided into two main sections; the first focuses on "personal and interpersonal effectiveness" and the second on "managerial and organizational development."

While some may find other management styles more to their liking, I have found Covey's principles to be particularly helpful for a technical communication manager. By its very nature, technical communication requires constant confrontation of ethical dilemmas and paradoxes. First, there is the ongoing conflict between quality and resources. Most technical communication managers must continually struggle with the paradox that they know how to produce better quality documents but lack the budget and time to do so. If those restrictions are too severe, managers may be forced to send out documents that are of little use to their audience. In the worst case, they may be asked to produce documents that put their audiences in danger or that cover up possible dangers. Each technical communication manager must decide how much he/she is willing to sacrifice in this trade-off, and where to put the proverbial sword in the sand and refuse to compromise any further, even if it means resigning. I once refused to change an employee performance review so that management-imposed quotas for "poor" performers could be met. I made it clear that I was willing to be fired, but that I would not falsify a document to meet personnel quotas. I was not fired, and the employee's rating stayed where I had put it. In another case, I realized that a document had improper instructions that could have led to physical danger for its users. I stopped shipment of several hundred thousand dollars worth of equipment until the documents could be fixed. The result was that the company did not meet its quarterly sales quotas, and several vice presidents were very irate. I was not fired (although it was threatened) and was in fact later told by the Vice President of Engineering that I had done the right thing. These two examples demonstrate the kinds of issues that face technical communication managers and that suggest that some kind of principle-based management philosophy is required.

There are, of course, many other types of management philosophies. Some managers pride themselves on being "bottom liners," on focusing on finances, budgets, and profits. Others focus on process more than on results, becoming enamored of the world of work breakdown structures, gantt charts, pert charts, deliverables, and milestones. Others are "people" managers, who work hard at being on warm, friendly terms with their own employees and everyone else. Others are "shippers," who take great pride in always making deadlines (sometimes to the detriment of quality). All of these philosophies can result in highly successful careers, especially if the environment in which one works fits the philosophy chosen. A good technical communication manager must constantly improve his/her skills in all of these areas. Some may buy into one particular philosophy so completely that the ethical dilemmas cease to exist. Shipping on time becomes so important that the quality of the documents does not matter. Most of us, though, must confront those dilemmas, and a principle-based management philosophy can help us do so.

Covey begins the section on managerial and organizational development by discussing seven chronic problems that he has observed in poorly managed organizations, as follows.

1. No shared vision and values: either the organization has no mission statement or there is no deep understanding of and commitment to the mission at all levels of the organization.
2. No strategic path: either the strategy is not well developed or it ineffectively expresses the mission statement and/or fails to meet the wants and needs and realities of the stream.
3. Poor alignment: bad alignment between structure and shared values, between vision and systems; the structure and systems of the organization poorly serve and reinforce the strategic paths.
4. Wrong style: the management philosophy is either incongruent with shared vision and values, or the style inconsistently embodies the vision and values of the mission statement.
5. Poor skills: style does not match skills, or managers lack the skills they need to use an appropriate style.
6. Low trust: staff has low trust, a depleted emotional bank account—and that low trust results in closed communication, little problem-solving and poor cooperation and teamwork.
7. No integrity: values do not equal habits; there is no correlation between what I value and believe, and what I do. (Covey 1990)

Unfortunately, many of us can recognize the problems, as we have seen them in full force at one or more of the organizations where we have worked.

Four Management Paradigms

Covey goes on to discuss four management paradigms. While they all have merit, he says, he believes that the first three are all based on flawed, limited perceptions of human nature. To get the best results, he believes that we must seek to operate on the fourth level, which leads to principle-centered leadership. Covey's first three paradigms align with historic trends in management philosophy. First, during the latter part of the nineteenth century and the beginning of the twentieth century, Taylor, Weber, and others developed the idea that organizations could be managed following scientific principles. As with the machines of the industrial age, organizations could be manipulated and controlled to produce greater efficiency and output. Then, around the first third of the twentieth century, partly in response to the inhumane philosophy of the scientific metaphor and the results it often produced, a human resource model began to be developed. This model accelerated after World War II, as nearly all organizations added human resource departments and began to develop systems designed to treat employees kindly and to recognize that efficiency and productivity were affected by how happy employees were. Finally, in the latter third of the twentieth century, when organizations began to realize the importance of employees' wishes to grow and to learn, they also began to stress the creation of learning environments and even to call themselves learning and knowledge companies (Morgan 1997). Covey describes these three stages using the metaphors of the stomach, heart, and mind, respectively. He also offers his own ideas for a more advanced metaphor based on the "whole" employee and using principle-centered leadership.

A brief paraphrase such as this can hardly do justice to two books devoted to the ideas behind a management philosophy. While Covey's business-oriented psychobabble can sometimes grate on a technical communicator, his underlying ideas are important enough to warrant consideration of his philosophies.

While different in its focus, *Managing By Values* (Blanchard and O'Connor 1997) proffers many of the same basic tenets. Although their book primarily treats organizational management issues, it also indirectly examines personal value systems. Blanchard and O'Connor describe a system wherein an organization chooses the basic, core values that it will use to define its success. The organization must then

TABLE 1.1	Covey's Four Management Paradigms		
Need	**Metaphor**	**Paradigm**	**Principle**
Physical/Economic	Stomach	Scientific Authoritarian	Fairness
Social/Emotional	Heart (benevolent authoritarian)	Human Relations	Kindness
Psychological	Mind	Human Resource	Use and Development of Talent
Spiritual	Spirit (whole person)	Principle-Centered Leadership	Meaning

work assiduously to align all parts of its operations with the value system. The organization's mission statement, longer-term goals, and shorter-term objectives must all be informed by its values. Blanchard and O'Connor describe a three-phase process establishing value-based management:

1. Phase 1: Clarifying our mission/purpose and values
2. Phase 2: Communicating our mission and values
3. Phase 3: Aligning our daily practices with our mission and values.

They describe in detail how each of these phases can be accomplished. The concepts they discuss seem relatively simple and obvious, yet it is easy to see that few organizations have successfully implemented them. Technical communicators may often have values that are not totally in accord with those of their larger organizations. While it is critically important for technical communication groups to align themselves with their larger organizations, it can also help to develop and maintain some of their own values, especially those related to ethics and to customer-based focus for the documents they produce. The concepts discussed by Blanchard and O'Connor will help show technical communication managers how to do so.

Mission and Vision Statements

Mission and vision statements are controversial. Scott Adams, the Dilbert cartoonist, has even included an automatic mission statement generator at his Web site: http://www.unitedmedia.com/comics/dilbert/career/bin/ms2.cgi. One simply chooses from a list of buzzwords, and the system then generates a suitable "mission" statement. The site rightfully pokes fun at the plethora of horrid mission statements that are, at best, collections of current "hot" terms communicating almost nothing and, at worst, outright lies.

However, mission statements can work very effectively for both large and small organizations. They provide a technical communication group with a means for defining itself and its goals within the larger context of the organization in which it works (Gilbert 1992; Eschen 1995). Often, the management structures and the organizational cultures define technical communication as a support service rather than a developmental one. Creating a mission statement for the technical communication group can help internally and externally. It can define the group as one that seeks to be more than a support function, that seeks to contribute more by creating products and services that meet the needs of customers more effectively. It can also communicate that same message to the larger organization, thus helping to move the communication group toward a more integrated, significant role in product/service development. Another benefit is that it will help technical communication employees understand that they are making meaningful contributions to the organization's outputs.

Creating a Mission Statement

Mission statements are more effective for a group when the group participates in developing them (Cook 1990). The technical communication manager should take advantage of the skills of the group members and engage them in helping to develop the mission statement. Following are ideas for creating an effective mission statement for a technical communication group.

1. Have the technical communication group participate in developing the mission statement.
2. Shorter is better. Long, overblown mission statements deserve the ridicule they get. Mission statements are more effective and memorable when they are short and snappy. Technical communicators are much more likely to respect a short, direct statement than one that is overblown with current buzzwords.
3. Focus on the results. The mission statement should emphasize outputs and results rather than the processes used to achieve them. For a technical communication group, the focus should be on the audience and on creating communications that meet their needs.
4. Motivate internally. A good mission statement should help inform the work of employees. It should give them an idea of the larger goals they are trying to achieve, even when they are developing something small.
5. Communicate externally. Ideally, the mission statement should communicate to the larger organization that the technical communication group understands and contributes to the overall organizational goals, while still emphasizing that technical communicators maintain a sense of fealty to their audiences. This means that mission statements should avoid terms that communicate the wrong message. If we want to be seen as more than a support group, we should omit the term 'support' or any of its synonyms. Rather, we should stress that we are developing products and services, just as engineers, programmers, and scientists do.
6. Communicate larger goals. The mission statement should imply goals beyond those that an organization can easily achieve in the short term. It should help stretch managers and employees toward constant assessment and improvement.

It is extremely difficult to achieve all of these things in a short, simple statement, especially if we want to make the statement work with items four, five and six above. Even for a group of professional communicators, constructing a statement that accomplishes all of these things can be a difficult challenge. Multiple one-hour meetings are a better way to accomplish this than is the forced, off-site, all-day meeting where the statement must be prepared by five o'clock, so that by four-thirty everyone starts giving up and putting in the latest management buzzwords.

The statement should, most importantly, resonate with the technical communicators for whom it will serve as a guideline. However, we should also look at the external communication purposes of the statement. One of its purposes is to help define the technical communication group to the organization as a whole in the manner in which it wants to be defined, not the manner in which it may have historically been defined, nor the manner in which less enlightened technical and scientific personnel sometimes define technical communicators.

Even if it cannot or will not be shared externally, developing a mission statement provides a technical communication group the means for defining how they conceptualize the work that they do, what they place value on, and what goals and aspirations they have. It can also help with formulating goals and objectives for longer and shorter time periods.

Searching the Internet using "mission statement" provides thousands of sites with examples of statements and suggestions for how to prepare them. Two places to start are the Bizplanit site at http://www.bizplanit.com and the Franklin/Covey site at http://www.franklincovey.com. For information on mission statements for non-profit organizations, see http://www.nonprofit-info.org/npofaq/03/21.html.

Developing Goals and Objectives

Once the mission statement is completed, it should be possible to use it as the basis for developing several long-term goals that are philosophical or strategic in nature. These goals expand the mission statement and also help to convert it from a concept into something that determines directions and actions for the group. The goals might express a long-term aim that the group eventually wants to achieve or they might express ongoing philosophies under which the group intends to work.

After the goals are developed, you can then develop objectives to go with each of them. The objectives are shorter-term aims that will define activities the group intends to complete quickly, usually within the current calendar or fiscal year. The objectives should be stated in terms that allow you to determine whether you have successfully completed them (Currie and Vallone 1995).

Table 1.2 provides an example of a mission statement (an uninspiring one, but offered here for illustrative purposes), a set of goals, and a set of objectives for each goal. Notice that the goals provide longer-term aspirations while the objectives cover the shorter term, usually the current year.

A mission statement and the related goals should be distributed to every employee who works in the organization. It is also a good idea to post them publicly in the halls and the conference rooms where the organization resides. The statement can also serve as a political document. It should be sent to the managers of all organizations with which the technical communication group works. This is especially true if the mission defines technical communication in a way that is more ambitious and more professional than its default definition within the organization. Sending out a new mission statement and goals will not change others' perceptions and misconceptions, but it is the beginning of the process. The real proof, of course, comes in making the mission statement and goals a part of everyday work processes. A technical communication manager must constantly consult the mission statement when making decisions. If those decisions run counter to the statement, then its value will be quickly lost. When employees see managers living by and pursuing missions, they will join in. When they see managers publishing mission statements that say one thing while management actions say something else entirely, morale and productivity quickly erode.

A mission statement and goals can help a technical communication organization begin to improve its processes and the quality of its documents. It can even

TABLE 1.2	Mission Statement, Goals, and Objectives

Mission Statement: To work together with other development organizations to design, prepare, and deliver the best documents in the industry.

Goal 1

Integrate technical communication with product development teams to achieve the highest quality products and documents possible

Objectives for Goal 1

Relocate technical communicators so they are within or adjacent to the development groups with whom they work, by June 1

Using the developers' Microsoft Project template, prepare a default technical communication template with tasks and milestone dates integrated into the standard development template, by April 1

Attend all project status meetings and present technical communication status related to schedule and budget for the project, by April 1

Goal 2

Develop a standard, consistent process for preparing product documents that excel within the industry

Objectives for Goal 2

Develop templates for technical communication information plans, project plans, and content specifications, by July 1

Develop forms for monitoring project status, including weekly time sheets, monthly time sheets, weekly reports, monthly reports, review submittal forms, change request forms, document control forms, printing specifications, and final reports, by September 1

Develop a customer satisfaction survey and send for all product releases, by July 1, receiving overall scores of at least 80%

Receive documentation scores of at least "very good" for products reviewed in our standard industry publication

Submit at least five documents to the STC awards competition, with at least two winning awards

Goal 3

Focus consistently on providing customers with the highest quality products that meet their needs

Objectives for Goal 3

For each project, develop a customer advisory team for documentation by April 1 and meet with that advisory team quarterly, in person, via teleconference, or online.

Working with the customer advisory team, perform audience and task analyses for each new release, with results to be included in the information plan for all projects after April 1

For each product, perform usability testing immediately after its release and issue a formal usability report for use on subsequent product releases, by September 1 contribute to improving the status of the group within the larger organization. As Covey (1990), Blanchard and O'Connor (1987), and many others have stressed, an organization with a strong mission that informs everything it does has a much greater chance of accomplishing short-term and long-term goals and of providing an enriching, rewarding environment in which to work.

Documentation Development Philosophy

A technical communication manager should not only develop an overall philosophy of management but also a philosophy of documentation development. Many technical communication managers do so, whether consciously or sub-consciously. Some are "shippers," who take great pride in always making deadlines, even if it might mean that some quality processes must be omitted to do so. Others are "aesthetes," who believe that the documents must be flawless in their technical accuracy and in the style and grammar with which they are written. Others are usability adherents, who believe that the documents must be based on extensive interaction and testing with the audience. Still others are "supporters," who see their role as supporting the scientists, engineers, or programmers with whom they work.

Most technical communication managers must balance these and other aspects of document development depending on numerous variables such as the time and budgets allotted for document development, the internal cultural and political realities surrounding how documents have been developed and are envisioned, the education and experience levels of the staff assigned to develop documents, relationships with management and with other groups, etc. In fact, to be effective, technical communication managers must learn how to succeed in meeting deadlines, ensuring accuracy, improving usability, and collaborating successfully with others. Regardless of their overall philosophy about document development, they must learn to accommodate all aspects of such development. Nonetheless, it helps a technical communication manager to have an overall philosophy that guides how documents are conceptualized and, especially, that helps guide decisions the manager must make when the inevitable conflicts arise among document development resources, timeliness, and quality.

Given all of the variables that affect how documents are conceptualized and developed in various types of organizations, it is impossible to proffer a single philosophy that will fit everywhere. However, I will offer my own philosophy, which is an admixture of radicalism and user-centered focus. Documents are products, and the people who develop them are product developers. They are not support personnel. As much as the labels on a piece of hardware (which are, after all, documents) or the user interface on a computer screen, documents are an integral part of a product. The success or failure of the overall product will be determined by how well its "engine" works, how well its user interface works, and how well its documents (in whatever form) work. Those pieces form an integrated whole that, from a customer's point of view, cannot be completely separated. Scientists may provide some of the contents for an environmental report, programmers may provide the code for a software product's "engine," and engineers may produce the circuitry for a radio, but the final products that go out to the various audiences provide an overall experience for those audiences that is an amalgam of the parts.

Hence, technical communicators should conceive of themselves as product developers, regardless of the nature of the documents that they create and of the current definitions of their work that others may have. A technical communication manager who conceptualizes the group's work as product development can and should fight for the same rights, privileges, and responsibilities as other product developers, even if it promises to be a protracted effort.

Secondly, when in doubt, technical communication managers should assume that they are user proponents and that they will do whatever they can to improve their audiences' chances of succeeding in the tasks they wish to accomplish. In many organizations, the technical communication group will be the only group with such a focus, as others will be more concerned about the science or technology with which they are working. Further, in the long term, technical communication groups that demonstrate they are improving audience experiences and that have the expertise and processes to rival those of other development groups are much more likely to achieve equality in status and in compensation.

Obviously, this philosophy will not work for everyone. Some, given the cultural, social, and political constraints of the organizations in which they work, will simply not be able to make claims similar to mine and/or will not care to engage in the polemics necessary to support the claim. Others will simply wish to work more collaboratively and less combatively. In any case, technical communication managers should consciously think about and develop their own document development philosophy to guide their decisions when they confront the inevitable dilemmas associated with tradeoffs among quality, resources, and project scope.

Organizational Structures

One of the most helpful ways to begin considering management philosophy in technical communication is to examine the various organizational structures that technical communication organizations can take. Starting with the question of where one will be organizationally and administratively can often aid in addressing other large issues related to how one will interact with others. Indeed, to an extent, one's management philosophy is often determined by organizational structures, even if it should be based on other considerations.

In typical corporate, government, and non-profit organizations, technical communications functions are organized in one of four basic ways. They can be physically centralized or physically dispersed and, within either framework, they can be administratively centralized or dispersed. While this subject has been treated often in the literature, it is beneficial to consider it again from the point of view of how the various structures influence the management philosophies of the technical communication managers.

TABLE 1.3	Comparison of Central and Dispersed Physical and Administrative Structures for Technical Communication Organizations	
	Physically Centralized	**Physically Dispersed**
Administratively Centralized	Pros: ■ Strong organizational identity ■ Greater consistency in style, standards ■ Flexibility in project assignments Cons: ■ Isolation from development groups ■ Support function status ■ Bunker mentality ■ Lower pay, status	Pros: ■ Closer contact with development groups ■ Greater knowledge of & identity with specific products ■ Perception as members of product team ■ Increased chance for audience interaction ■ Higher pay, status Cons: ■ Lower identity with central TC organization ■ Less consistency in styles, standards ■ Isolation from other TC employees ■ Reduction in centralized boilerplate, solutions
Administratively Dispersed	This model is rare. Pros: ■ Lower management costs Cons: ■ Isolation from managers physically & professionally ■ Lower pay, status	Pros: ■ Lower management costs ■ TC'ers full members of product team ■ Strong knowledge of & identity with product and team ■ Increased chance for audience interaction Cons: ■ Risk for TC during layoffs, cutbacks ■ Less consistency in styles, standards ■ Isolation from other TC employees ■ Reduction in centralized boilerplate, solutions

Overall organizational structure affects the structure of communication groups in two primary ways: administratively and physically. The administrative issue boils down to whether there is a central technical communication organization. If so, the technical communicators report to their own organization. If not, they report directly to managers in other organizations, usually in engineering or marketing, depending on the nature of the overall organization. The physical issue addresses where the technical communicators work, either in a central place or dispersed among project groups throughout the organization. Hence, the administrative/physical arrangements shown in Table 1.3 are possible. Obviously, many more pros and cons exist for each of these possibilities, but I have listed only the most significant here.

Central Administration, Central Location

The administratively and physically centered model gives communicators the greatest identity with each other as a group. They have immediate access to peers, which means they can run ideas past like-minded people, get help with questions about development tools, communicate quickly about what others have done on similar projects, provide mentoring and coaching for one another, etc. Unfortunately, this model also encourages a kind of "bunker" mentality, with the communicators bonding together against the semi-literate technicians with whom they are forced to work. Even worse is the way in which such arrangements are viewed by others in the organization who tend to see the group as a roadblock and an extravagance that provides "wordsmithing" services but little else of value. As Plung (1994) puts it, such a centralized group is often viewed as "…an enclave of ·dilettantish sophists who have no real comprehension of the precise nature of the company's mission or its technical communications." The physical and administrative isolation allows communicators to reward each other and ignore the organization's overall missions and goals.

This organization follows the industrial age concept that organizations should be structured according to functions. It is a conservative approach based on the concept that ideas and products are developed through a kind of assembly line structure, with experts in each required function located in the appropriate places on the line. It usually requires the waterfall method of development, a highly linear system in which the product or service is handed from one function to another at appropriate times.

The management philosophy in such an environment is usually a conservative one, focused on top-down directives and monitoring. In the information age, this is a difficult structure for technical communicators to work in, as information flow tends to be hierarchical and linear, making it difficult for communicators to add value to the organization's outputs. Frequently, engineering specs or requirement documents are handed to the communicators to "pretty up" or "wordsmith" with the expectation that they will not make significant changes to the structure or content of the material. While there are exceptions in organizations with enlightened management, this structure requires constant striving on the part of a communications manager to integrate the communications function into the rest of the organization.

Dispersed Administration, Dispersed Locations

The organizational model at the other extreme essentially eliminates any centralized technical communication organization, and communicators report directly to the product/project team managers with whom they work. This system can work quite effectively with the right managers. However, technical managers tend to value the work of technical people more highly than that of communicators. Given limited resources for raises and promotions, those managers will distribute them first to the employees they see as more valuable, the engineers and programmers. While communicators in such groups may have a stronger identity with their products and with the development team, they are likely to suffer from lower remuneration and from very limited career prospects. Further, many of them have to re-invent the wheel on each new project because the organizational structure discourages centralized standards in document design, style, layout, file structures, tool usage, etc.

Management philosophies differ radically, from "clueless" to enlightened, among managers who have not been technical communicators. Obviously, the job satisfaction and the fate of the communicators depend on where their particular manager falls on that continuum. Under the right circumstances, this can be a very rewarding system for a technical communicator who is a valuable member of the team and contributes to the overall design and quality of its products. In the wrong circumstances, it can be a nightmare, leading to few rewards and low self-esteem, with the resulting high burnout and turnover rates.

The overall philosophy behind the dispersed structure is that projects are more important than functions, and that for projects to succeed they must have all of the functions working together administratively and physically. Indeed, many managers in such a system encourage the de-functionalizing of job roles, so that the project receives the benefits of cross-fertilization among the various team members as they assist one another in achieving project goals. In the information age, this model works better than the more centralized systems because it allows information to be shared quickly among all team functions and encourages cross-functional collaboration and problem solving.

Dispersed Administration, Central Location

This model is very rare. It makes little sense to have a centrally-located group managed by people not associated directly with that group. Further, because the communicators are isolated from their management, they are unlikely to receive adequate management direction and support. I have seen cases where an on-site project manager at a customer's site directed the efforts of a central group at a contractor's headquarters elsewhere. With enhanced communications through e-mail, voice mail, on-line video, etc., such an arrangement can work for project-related activities. However, while the remote manager can successfully perform project management activities, it is considerably more difficult to perform personnel hiring, assessment, and evaluation for a group that is entirely remote. It is very difficult for a remote manager to sense group morale and to tell if communicators are dedicated to the project or are merely going through the motions. Such arrangements seem to work better when the project manager is able to spend some "face time" with the communicators, preferably at regularly scheduled project progress meetings.

The philosophy behind this model assumes that work can be done efficiently, without requiring communicators to be co-located with the project. There are considerable cost savings on a contracted job if only one person works at the client location and the remainder of the project team stays at the contractor's offices.

However, for the model to succeed, special management skills are necessary. The on-site project manager must have superb communications skills, both with the clients and with the project team back home. The manager must know how to motivate people to meet deadlines with quality products, even though the manager is not monitoring their daily activities. The manager must also have excellent project management skills, and must communicate project progress, deadlines, milestones, and problems effectively and promptly. In other words, this model requires a manager with a full breadth of management expertise and abilities.

As more communication projects are done on an outsourced basis (Web page development, for example), this model is likely to become more prevalent, assuming that there are enough good technical communications managers to fill the required positions.

Central Administration, Dispersed Locations

This model offers many advantages and few disadvantages. Its primary advantage is that it gets communicators in the same physical space as technical and scientific personnel, so that they are much more likely to be seen as equal team members. It also means that communicators are there when a new release or revision is started, so that they work on it from the very outset. The model has the further advantage that communicators are managed and rewarded by management familiar with the work of technical communication rather than by people in engineering or marketing who may not understand exactly what communicators do or how they add value. This arrangement's biggest disadvantage, that it is harder to create standards and use boilerplate solutions for higher productivity, can still be overcome with good management practices. As Killingsworth and Jones (1989) point out, this model leads less to a division of labor philosophy and more to an integrative philosophy in which communicators work directly with other functions and become contributors who are more psychologically and financially satisfied.

This model also leads to what is often called matrix management. In this case, the dispersed communicators report on a daily basis to the technical managers whose groups they work with, while still reporting administratively to a central technical communication manager. Such a system satisfies the technical manager, who remains in control of all aspects of the project, but it also avoids the problem of the technical manager directly managing someone whose discipline the manager may know little about. However, the communication manager may have to rely heavily on the technical manager's assessment of the employee's work, which can lead to some uncomfortable political and personnel problems.

Despite its few difficulties, the centrally administered, dispersed location model is popular with contracting firms and, increasingly, within larger corporate and government organizations. It provides technical management with control over the day-to-day work of more of their project team but avoids the burden of having to hire and supervise people whose jobs they are not familiar with. It gives the technical communicators the opportunity to work closely with technical personnel, but to continue to be members of a central group of other communicators. They can look to that group for help with problems and for productivity advantages resulting from standard style guides, templates, etc. Technical communication management has considerable flexibility with this model, as they can move people around as projects ramp up (and down), and they can offer their employees continued employment and variety in project assignments, rather than going through the hire-and-layoff cycle when a project ends.

The philosophy behind this model is that technical communication is an important enough part of a project that the practitioners should be located with others on the project. It also assumes that there are advantages, both in productivity and in morale, to having a separate, centrally administered technical communication organization.

Flattened Organizations

The industrial age model of hierarchical management structure does not work well in the information age. Indeed, except for extreme cases, like war, it was probably not the best model even for industrialists. In an organization where everyone has access to information at the same speed, the hierarchical model becomes slow, inefficient, and expensive. Because of this, many organizations are working to eliminate levels of management and to configure new structures that bring greater efficiencies in developing products and services.

These new structures involve increased reliance on project teams, parallel communications systems rather than serial, hierarchical systems, emphasis on results rather than on processes, emphasis on employees' skills and abilities rather than on static job definitions, and a recognition that an organization should strive for and embrace constant change rather than attempting to achieve a peaceful stasis (Zuboff 1989).

What does this mean for technical communicators? In terms of the four structural models discussed previously, it suggests that less central administration is more desirable and that dispersed project teams are more desirable than centrally located ones. Employees in organizations pay more attention to management behavior than to management pronouncements. No matter how much managers say they want to change an organization's culture, if they keep in place highly centralized, hierarchical structures, they are not going to successfully flatten their structure and meet the requirements of the effective information age workplace.

Technical communication managers should aim for just enough central management to gain its advantages (templates, boilerplate, design standards, etc.) while allowing the optimal amount of administrative and physical dispersion necessary for effective project teams.

In general, this means larger groups of direct reports to a single manager. In hierarchical systems, it was assumed that a manager could not direct the activities of more than about eight to ten reports. In the flattened workplace, managers have as many as fifty direct reports, which means, of course, that they cannot begin to have the same level of oversight over individual employees. They must cede power to those employees, including even such matters as performance assessments and reviews, which must now be conducted collaboratively and not merely delivered from the manager to the employee.

Where Should Technical Communication be Located?

Another basic philosophical problem concerning technical communication groups is where they should be located. Should they be in the engineering or programming organization? In the sales/marketing organization? In a separate organization? To whom should they report?

The administrative location of a technical communication group often says much about that organization's attitudes toward the value and importance of good documentation (Wishbow 1999). In one organization where I worked, the technical communication group was part of support services, along with the cafeteria, warehouse, and janitorial staff. Indeed, its reporting structure indicated its importance within the overall organization. It also indicated how many resources were dedicated to technical communication and what the subsequent quality of the documents was.

There is no simple, universal answer as to the most appropriate location for a technical communication group. There are too many variables, and individual organizations' structures and cultures vary too widely to make any sweeping generalizations. However, there is a single, overriding consideration that should affect where technical communication groups belong. In the information age, the outputs from a technical communication group are *not a support service*. They are a crucial component of the products and services with which they are associated. They are as much a part of the product as any bolt or nut or line of software code. In fact, if they are included in an on-line system, they are indeed a part of the code for a software product. If a technical communication group develops a company's Web site, they are not providing a mere support service. From the customer's viewpoint, the Web site is the company. Following this principle, it makes sense to locate the technical communication group within the organization that develops the various products and services.

In a software house, the communication group would be located in the programming organization. In a hardware company, the group would be located in the research and development (R&D) engineering organization. A technical communication group responsible for proposals would be located within the sales organization. A group responsible for filing pharmaceutical reports to the FDA would be within the R&D organization.

Many technical communicators do not want to be located within groups largely composed of scientists or engineers because they have experienced a severe case of "second-class citizen" status within such groups. They are classified lower, paid less, and laid off first in organizations that value scientific or engineering expertise over making products and services more usable for customers. They suffer from what Lesandrini (1999) calls the "Bottom of the Pile Syndrome." In some organizations, the scientific/engineering culture is so strong that even a technical communication manager's most arduous education efforts are not going to change it. In those cases, it is probably better if the group is located elsewhere. However, in many cases a campaign by technical communication management can help demonstrate how important the group's outputs are to the success of the organization. (See the sections on Value Added and Public Relations in Chapter 3.) Technical communication managers should constantly seek ways to demonstrate value and to improve the status of the technical communication group, regardless of its physical and administrative location.

Perhaps the ideal is the centrally administered, physically dispersed model discussed earlier in this chapter. Here, the communicators are co-located with the product and service groups with which they work, but are administered by a central technical communication organization more sympathetic to their career goals and more capable of enhancing efficiency.

Conclusion

Regardless of where they are located physically or administratively, technical communication managers need to consider the philosophical issues associated with managing their groups. Whether through formally developed missions and goals or by other means, technical communication groups should develop larger goals than merely getting the job done. Because technical communication groups often do work that differs from others in their organizations, it is particularly important for them to define who they are and what they want to accomplish. It is also important that they communicate their goals and their methods throughout their organizations.

By its very nature, technical communication involves complex and varied ethical dilemmas. Perhaps the most persistent is the conflict between quality and available resources. Confronting such dilemmas is unavoidable, but such confrontations are considerably more manageable when one has a set of management principles as a guide. Principle-based management can help with decisions about where technical communication groups should be within an organization, both physically and administratively. While it is important for a group of technical communicators to contribute to the overall goals of the larger organization, it is also important that the communicators be in a position to be regarded, treated, and compensated as professionals.

Another advantage to having principles and values inform one's management style is that they help a manager interact successfully with the people who are direct reports. The next chapter covers personnel management, including those aspects of it that, in the technical communication field, differ from others.

References

Abrahams, J. 1999. *The mission statement book: 301 corporate mission statements from America's top companies.* Berkeley, CA: Ten Speed Press.

Blanchard, K., and M. O'Connor. 1997. *Managing by values.* San Francisco: Berrett-Koehler.

Cook, K. J. 1990. How strategic planning can work in your organization. *Technical Communication* 37(4), 381–385.

Covey, S. R. 1989. *The 7 habits of highly effective people.* New York: Simon & Schuster.

————1990. *Principle-centered leadership*. New York: Fireside.

Dombrowski, P. 2000. *Ethics in technical communication*. Boston: Allyn and Bacon.

Eschen, M. L. 1995. "Keeping our sanity: managing change in an ever-changing world." In *STC 42nd Annual Conference Proceedings*. Society for Technical Communication, 220.

Gilbert, C. E. 1992. "Managing in the 90's: Vision for the future." In *STC 39th Annual Conference Proceedings*. Society for Technical Communication, 730–732.

Killingsworth, M. J. and B. G. Jones. 1989. Division of labor or integrated teams: A crux in the management of technical communication? *Technical Communication* 36(3), 210–221.

Lesandrini, J. 1999. "Pulling yourself up by your bootstraps: Strategies for advancing your documentation team's position in your company." In *Conference Proceedings: 17th Annual International Conference on Computer Documentation*. New Orleans: Association for Computing Machinery (ACM) Special Interest Group on Systems Documentation (SIGDOC).

Morgan, G. 1997. *Images of organization*. Thousand Oaks, CA: Sage.

O'Hallaron, R. and D. O'Hallaron. 2000. *The mission primer: Four steps to an effective mission statement*. Richmond, VA: Mission Incorporated.

Plung, D. L. 1994. Comprehending and aligning professionals and publications organizations. In *Publications management: Essays for professional communicators*. Edited by O. J. Allen and L. H. Deming. Amityville, NY: Baywood 41–54.

Rude, C. 1994. Managing publications according to legal and ethical standards. In *Publications management: Essays for professional communicators*. Edited by O. J. Allen and L. H. Deming. Amityville, NY: Baywood 171–187.

Wishbow, N. 1999. Home sweet home: Where do technical communication departments belong? *Journal of Computer Documentation* 23(1): 28–34.

Zuboff, S. 1989. *In the age of the smart machine: The future of work and power*. New York: Basic Books.

For Further Information

The "classics" of principle-based management are Stephen R. Covey's *Principle-centered Leadership* (1990) and *The 7 Habits of Highly Effective People* (1989). These books provide not only the conceptual basis for principle-based management but also highly pragmatic information about how to practice it. A different conceptualization is provided by Blanchard in *Managing by Values* (1997). While Covey suggests that leaders should start by analyzing and expressing their own personal principles before considering leadership principles, Blanchard proposes that organizations should select their overriding values, state them, and live by them.

The books mentioned above all discuss the importance of having effective organizational missions and using them to develop goals and objectives. Jeffrey Abrahams's *The Mission Statement Book: 301 Corporate Mission Statements from America's Top Corporations* (1999) provides historical information, copious examples, and a section on preparing statements. O'Halloran and O'Halloran's *The Mission Primer: Four Steps to an Effective Mission Statement* (2000) provides a more concise, pragmatic guide to creating mission statements.

Of the plethora of books that cover organizational structures, Gareth Morgan's *Images of Organization* (1997) provides an excellent summary of each of the various metaphors that have been proposed to describe such structures, including the scientific or machine metaphor, the organism, the brain, culture, political systems, psychic prisons, chaos and change, and domination. Shoshana Zuboff's *In the Age of the Smart Machine: The Future of Work and Power* (1989) discusses organizational structures and the effects on them of computers, technology, and the information age. Daniel L. Plung provides valuable ideas for fitting technical communications properly within larger organizations in his chapter, "Comprehending and Aligning Professionals and Publications Organizations" (1994).

Nina Wishbow's article, "Home Sweet Home: Where Do Technical Communication Departments Belong?" (1999) discusses the advantages and disadvantages of locating technical communication within other departments, including marketing, product line management, and development or engineering.

While not treating principle-based management directly, much of the literature on ethics in technical communication treats it at least indirectly. Paul Dombrowski's *Ethics in Technical Communication* (2000) touches on principle-based management in numerous ways, as does Carolyn Rude's chapter, "Managing Publications According to Legal and Ethical Standards" (1994).

Questions for Discussion

1. Is it possible to use the concepts of principles and values related to management? Aren't managers ultimately responsible for making profits? How can you reconcile the idea of principle-centered management with making profits?

2. Consider the depictions of management people that you see in the popular media, especially television programs and movies. Are managers generally treated as principled people? Why do you think they are depicted as they are?

3. Find one or two examples of mission statements that you believe to be especially effective. Likewise, find one or two examples of statements that you find to be especially ineffective.

4. Mission statements have been around since the nineteenth century. Are they obsolete relics of the past when autocratic, hierarchical management structures prevailed? Or are they valuable tools that can energize a group or an entire organization despite the management structure?

5. During the 1980s, management by objectives (MBO) became one of the management "fads." Case 1 requires you to use a form of MBO. Have you used this method before? Have you found it to be valuable or not? Give some examples.

6. Have you ever been given a written set of objectives to work toward during a given year? Did you prefer to work with such structure or would you prefer a more freewheeling approach? If you are not given written objectives, what other methods are used to provide guidance about what is expected of you?

7. Of the four organizational structures for technical communication organizations, which one would you prefer to work under? Why?
8. Where should a technical communication group be placed administratively? In sales/marketing? Engineering? Customer support? Its own separate organization? Why?

CASE 1

Mission Statement, Goals, and Objectives

The Management Situation

Aardvark Enterprises has decided to enlarge its technical communication group, with you as the new group manager. Congratulations on your promotion!

Aardvark is a small company with about 200 employees, 120 of whom are engineers and programmers. Aardvark designs and develops both hardware and software products to support point-of-sale systems, a highly competitive niche market. A secondary product line is financial analysis software. Aardvark's products have consistently been better designed than competitors' products, but customers have reported frustrations with learning how to use them, and returns have been high.

Your first assignment from upper management is to prepare a mission statement, goals, and objectives for your new organization. You face several challenges in creating these documents because they will have to work with multiple audiences.

Aardvark's Engineering Department has a somewhat conventional view of technical communication, that writing manuals and online help is something done by people who can't hack it as programmers, who are not "techies." Before you were hired, most of the documents were written by entry-level engineers who were learning the ropes or by discredited engineers who were being moved off technical assignments and toward the door.

Your new boss is an enlightened Engineering Manager who believes in the importance of good product documentation. Likewise, his boss, the VP of Engineering, believes in its importance. In fact, it is she who pushed to get you hired and to set up a permanent technical communication group. These are the two people who have given you the assignment to create the documents. They are happy to have you on board and are expecting great things from you. They also have made it clear that they will help you get started and will, at least to an extent, offer support (psychological, if not financial). However, they expect you to show leadership in setting up your group and making it an important part of the engineering development community.

So far, you do not have any employees working for you, but you have projects that are going to require hiring at least three full-time people in the next six months. Those employees will work in your group and will be informed by the mission, goals, and objectives you prepare. They will have to accept them and be motivated by them if you are going to be successful. If you write unimaginative statements of the sort attacked by Dilbert and others, you will lose credibility with your new employees.

Your mission, goal, and objective statements will have to make each of those groups happy. You need to gain credibility within a skeptical Engineering Department (management support notwithstanding). You want the engineering groups to regard your people as a peer group, one that contributes its fair share to the products that Aardvark makes. You want management to see that they made the right choice in promoting you and that you are on the right track toward setting up the type of organization they are looking for. And you must also prepare the three statements so that they will challenge and motivate your future employees.

The Assignment

Prepare a mission statement, a set of goals, and a set of objectives for your new group. The mission statement should be brief, succinct, and have some punch. It should provide members of your group with a vision of what your organization will be and how it will work.

The goals should be long-term goals, things that you want to accomplish in the next two to four years. The goals should also include ongoing aims of the group that will be on the list every year, like "preparing documents that are the best in the industry."

The objectives should be for the short-term, things you hope to accomplish in this calendar year. Each objective should be tied to one or more of the goals, and they should all contribute to achieving the mission.

Your report should be no longer than three pages, with the mission statement, three to five goals, and five to ten objectives. The goals can be stated in general terms, without specific measures of success attached to them. The objectives, on the other hand, should have specific measures of success. For example, you might satisfy a goal of achieving customer satisfaction with an objective that says that you will receive scores of at least 90 percent approval on every document for which you survey your customers.

Helpful Hints

Try some of the following sources for more information about missions, goals, and objectives:

Abrahams, Jeffrey. 1995. *The mission statement book: 301 corporate mission statements from America's top companies.* Ten Speed Press.

Brinckerhoff, Peter C. 1994. *Mission-based management: Leading your not-for-profit into the 21st century.* Alpine Guild.

Jones, Laurie Beth. 1996. *The path: Creating your mission statement for work and for life.* Hyperion.

Jones, Patricia & Larry Kahaner 1995. *Say it & live it: 50 corporate mission statements that hit the mark.* Currency/Doubleday.

http://www.bizplanit.com

http://www.franklincovey.com

http://www.businessplansoftware.org/advice_mission.asp/

http://www.nonprofits.org/npofaq/03/21.html/
http://www.quintcareers.com/mission_statements.html
http://www.tgci.com/publications/98fall/MissionStatement.html

Evaluation Criteria for Case 1

Does your mission statement fit the needs of each of the audiences that will see it? Is it succinct and meaningful? Could it have been created on the Dilbert automatic mission statement creation page? (If so, upgrade it.)

Do your goals give each of the audiences a long-term vision of what you hope to accomplish?

Do your objectives give employees a clear idea of what you expect from them in the current work year?

Do the three documents work together as an integrated whole?

CHAPTER 2

Technical Communication Personnel Management

Introduction

Technical communication is an enterprise that requires human effort but little capital investment or equipment. For this reason, the personnel management abilities of technical communication managers are especially important. The success (or failure) of a technical communication organization requires managers who understand how technical communicators work and how their creations add value to the organization in which they work. Those managers must convert that understanding into appropriate behavior that keeps technical communicators motivated, rewarded, and learning.

In a fairly common scenario, technical communication is lodged in an engineering, programming, or scientific environment where the manager in charge does not understand or appreciate what the technical communication employees do or what value they add. In such situations, morale is often low, turnover is high, and no one is happy with the way technical communication gets done. A strong, knowledgeable manager makes a tremendous difference in how well technical communicators work and in how they are perceived by others.

Good personnel management skills are the most important part of a technical publications manager's repertoire. After discussing leadership and management, this chapter will discuss the personnel management tasks that technical communication managers must perform, from considering how to staff a group to hiring employees, mentoring them, evaluating them, rewarding them, and, when necessary, removing them. It will then discuss other personnel-related matters, including team building and collaboration, telecommuting, and personal leave and day care.

Leadership vs. Management

Management and business publications are filled with articles about leadership. Indeed, there is an entire industry devoted to teaching people in management positions how to be leaders. While I will not try here to summarize or reproduce all of the ideas, I will try to differentiate between managing and leading, and to proffer the idea that for technical communicators certain leadership styles work more successfully than others.

A manager is someone who takes care of business, including balancing the budget, making deliveries on time, handling all of the paperwork required, and making sure that an enterprise meets its immediate goals. A leader, on the other hand, strives to realize longer-term visions and to inspire fellow employees so that they meet both short- and long-term goals. As Aughey (1999) puts it, "Managers do things right. Leaders do the right things. Both roles are necessary, but they differ in style and result." Many leaders delegate as many of the management tasks as they can, so that they can focus their energies on planning, developing, and implementing long-term goals. Hence, in my definition, it is possible for someone to be a superb manager and a poor leader, and vice versa. The most successful management people are those who are able to do both—to visualize long-term aspirations and to inspire others to want to reach them, while simultaneously keeping track of the immediate details and events necessary to reach the ultimate goals.

Good technical communication management requires both skill sets. As Carlson (1999) puts it, "Put simply, modern expectations require that a manager be able to lead people as much as manage resources." In many organizations, project management skills are so important that communicators who show the best project management skills are the ones who are promoted into management positions. Being able to meet immediate deliverable dates is of paramount importance. The danger, though, is that such an organization lacks real leadership; there is no one looking to the future and planning for the inevitable changes in tools and processes to be used for developing documents and on-line systems. Writers who work for project-management oriented supervisors often enjoy knowing exactly what the status of the project is, but they also often burn out after months and years of getting little information about what their next assignment will be, what their growth path is, or how they fit into the larger organizational picture.

A good technical communication supervisor must have project management skills, as missing deadlines and routinely going over budget are simply unacceptable in most organizations. However, those skills are a necessary but not sufficient condition for achieving success as a technical communication supervisor.

Excellent leadership skills are also necessary. Following are the most important leadership skills for someone working with technical communicators.

Leadership Skills for Technical Communicators

Give Them the Big Picture

Leaders show people contexts, which can range from how a document fits into a specific product set all the way to how a company and its products fit into and contribute to world society (Horowitz 1994). It is important for technical communication leaders to provide both the smaller contexts (what a writer is doing on a specific project and how it contributes overall) and the larger contexts (how the organization and its mission fit into larger societal considerations).

This activity should start early and should be ongoing. Showing job candidates both the larger context of the organization and the particular group/project on which they would be working is a good recruiting technique and a useful method for assessing the cultural match between candidate and organization. Draw pictures on a whiteboard or a piece of paper to show the candidate where the job responsibilities fit. And do the same thing with existing employees on a regular basis. Each time you give an employee a new assignment, provide the context for how it furthers the organization's goals and missions (Hansen 1988). Technical communicators often see only a tangential link between what they are doing and the organization's overall goals. It is important to show them how their assignments fit into the larger organizational context.

Give Them a Vision

When you give an assignment, you should provide the communicator with your vision as to what the optimum outcome should be. Depending on the nature of the project, this may involve a vision of both the content and the physical appearance of the finished product. Simply saying, "Go write a 50-page user guide," is not enough. The communicator should understand the full context of the document's use and should have some idea as to your vision about what the document will look like.

I once gave an employee a two-thousand–page user manual (if you can imagine that!) covering three pieces of equipment. The document, needless to say, was laden with verbosity, repetition, and redundancy. I explained my vision that the document should be three two-hundred to three-hundred-page user guides, one for each piece of equipment that provided only the information needed by a user of that model. I elaborated about organization, access methods, etc., without going into so much detail that the employee wouldn't have some thinking and designing to do. I further stated that I thought we ought to be able to achieve the vision within two releases of the document. The employee got excited and began immediately to disassemble the monstrosity. In two releases we indeed had our three shorter, more effective (and economical) documents, and the grudging admission of engineers that maybe sometimes less is more.

Another important part of providing vision is to explain clearly expectations for quality. Employees can get very excited about the possibility of creating something of the highest quality, something that will win Society for Technical Communication (STC) awards for design and content. However, if all you want (or can afford) is a simple revision without any improvement in quality, you should make this clear right from the start. In this case the vision is limited, but employees need to know what is expected when they get a new assignment.

Likewise, if you expect someone to depart from the traditional way of doing things in your organization or industry, you need to provide them a vision of the new outcome that you hope for. Provide as many specifics as possible concerning both the content and the design of the new information.

The need for giving employees a vision is explained, in part, by expectancy theory (McGuire 1991). Expectancy theory says that the likelihood of someone acting in a specific manner increases with their expectation that such acts will lead to a given outcome and further increases as the value to them of that outcome increases. Hence, three variables become important: (1) the attractiveness that the employee places on the outcome, (2) the performance-reward linkage that the employee sees between acting in the desired manner and the rewards to be received for doing so (usually both psychological and financial), and (3) the effort-performance linkage the employee sees between the likelihood that exerting effort will lead to the desired performance and thus the desired outcomes. Employees who understand the final outcomes desired and the rewards that will accompany them are more motivated to perform productively.

Give Them a Path

Employees want to see a path for their organizational and career advancement. They want to know how their current assignment contributes to the opportunity for growth and advancement within the organization, and how it contributes to their overall career growth. Leaders try to communicate such information as often as possible, at the very least at the outset of each assignment. It is a good idea to have at least one formal career path discussion with each employee each year, to map out their short- and long-term goals and the means by which they can achieve them. A technical communicator needs to see how a current assignment will contribute to their moving up one salary band or one rank or both. More than once, I have given communicators an assignments and told them that if they successfully completed the assignment and did so at an above-average level, that I would promote them.

In smaller organizations there may not be any apparent career path for a technical communicator beyond moving up through the various levels at which technical communicators are classified. If communicators have moved into positions in other departments, a technical communication manager can explain to direct reports that they have options for career growth outside of technical communication. Such opportunities can come in management, corporate communications, advertising, marketing, sales, customer relations, etc.

Defend Your Organization (Not Your Craft)

This entry has more to do with leadership within an organization than it does with leadership of employees. Technical communication managers have often made the mistake of defending the craft of writing. As Schriver (1997) points out, this view of technical communication as craft led to the once widespread notion that technical writers were little more than glorified "wordsmiths" and secretaries, and that they made minimal contributions to the real value of an organization's products and services.

What technical communication leaders should emphasize, instead, are the contributions that their organizations make to the larger organizations' goals, even if those goals are unashamedly financial. Technical communication leaders who defend the craft put themselves and their organizations in the role of gatekeepers and roadblocks rather than as contributors, thus making themselves highly expendable. True leaders show their organizations (and their employees) how they can help contribute to and add value to the products and services that the overall organization creates. In one of the organizations where I worked, technical communicators achieved equality in pay and rank with the technologists, but it required de-emphasizing our "craft" and emphasizing instead the unique processes and capabilities that our jobs required. It also required joining more closely with the development groups in working collaboratively on projects.

Tailor Communication to Individuals/Situations

Leaders understand that different people and groups require very different communication methods in different situations (McGuire 1991; Muench 1995). This is discussed in more detail in the chapter on Communication, but it is important to note that leaders adapt their communication modes to particular people and situations. Some employees need constant communication from a supervisor, while others prefer to be given an assignment and then be left alone. The smart leader learns (sometimes even asks) about the preferences of employees and adapts to them. Smart leaders do the same with peers and with their own supervisors. Technical communicators are more likely than most employees to be especially attuned to both the style and content of communications that they receive. Hence, technical communication managers must work more diligently than others to ensure that they are communicating with each person in the most effective manner possible. This requires conscious analysis of each person's communication preferences and the use of appropriate communications methods to meet those preferences.

Develop Strong Peer Relationships

Even though technical communication managers often have little in common with other managers within an organization, they should still try to develop good working relationships with those peers. This has obvious benefits for the success of a technical communication group, and employees are much more likely to follow someone whom they perceive to be trusted and respected by other managers. Many technical communication managers make a mistake when they isolate themselves physically and psychologically from their peers in technical organizations. The bunker mentality that results can only detract from the view that the technical communication organization is an equal participant in creating the important products and services that the organization delivers. Further, it presents the managers' image and the image of the technical communication organization as one that is different, less important, and less essential. Many new technical communication managers face a significant challenge in upgrading perceptions both of their overall organizations and of their employees, due to the bunker mentality of a previous technical communication manager. One of the fastest ways to start to improve things is to develop good relationships with managers in other groups.

Believe in What You're Doing or Don't Do It

This is one of the major paradoxes for a leader. What do you do if upper management passes down a new edict or policy and requires you to enthusiastically disseminate the information to your group? Do you fake enthusiasm even if you disagree with the policy? Or do you disobey the order by either not disseminating the information or by doing so in a way that makes it clear that you disagree with it? Technical communicators are people whose profession is communicating information. They are going to know from your rhetoric and even your body language if you are trying to fake enthusiasm for something. If you cannot agree with organizational policies, it is better to explain to your employees which aspects you embrace and which you disagree with, and why. You risk getting in trouble with your supervisors, but you maintain your integrity in the eyes of the people who report to you, and no one is going to follow a leader whom they believe to lack integrity.

Help Everyone Who Wants to Leave

People follow leaders whom they perceive to have their best interests at heart. If they see a leader prevent people from getting promoted or from transferring to another organization where they can learn and grow in a new position, they will assume that the leader has his/her own best interests at heart rather than those of the employees. One goal of a leader is to help every person in the organization to achieve to the best of their ability. If that can happen in another internal organization or in an external organization, help make it happen. My philosophy is that it is better to have ex-technical communication people in as many internal organizations as possible, especially if they have been promoted into management positions (Maggiore 1991). If employees see a bitter attitude from a manager toward someone who has given two-weeks' notice, they will assume that the manager would do the same thing to them. Help everyone get ahead, even if it means short-term problems for you in quickly finding and hiring replacements. The long-term rewards, especially concerning the perception of one's leadership, outweigh the immediate problems.

Get Out of the Way and Get Other Things Out of the Way

If a leader gives someone a clear picture of the context of a particular assignment and then provides a vision of the desired outcome, but does nothing to help the employee get there, the leader loses all credibility. It is important for a leader to allow employees to arrive at solutions on their own, without micro-managing every detail of the effort. However, it is also important for a leader to become involved, when necessary, in removing roadblocks and impediments to the employee's chance for achieving the vision. This requires striking a fine balance between interceding too often and not often enough. Further, different employees will want and need different levels of involvement from their supervisor. Trying to use the same method of interaction with everyone does not work, so a manager must figure out for each of their direct reports how to help when necessary and how to stay away when necessary. One of the most common project-related difficulties for technical communicators involves problems with subject matter experts (SMEs). The communicators might not have enough access to SMEs to get required information, or they may be getting too much information, accompanied by demands that it all must go in the documents. In either case, a manager must intercede on the behalf of the writer to get things cleared up. This is another reason why good peer relationships are so important. Likewise, a technical communication manager must intercede quickly and decisively when software or hardware problems are impeding communicators from working as effectively as possible.

Keep Reading, Learning, and Thinking about Leadership

This brief section can only treat a few ideas on how to improve as a technical communication leader. Good leaders never stop learning and thinking about leadership and how they can be better at it. It is important to continue to refresh one's thinking by reading articles and books, taking seminars and courses, having discussions with peers and supervisors, and doing everything possible to continue to grow as a leader. See, for example, the American Management Association (http://www.amanet.org/index.htm) and Franklin/Covey courses on leadership (http://www.franklincovey.com).

Leadership is increasingly difficult and complex in the transition from an industry-based economy to one that is information based. As people who develop, design, and disseminate information, technical communicators have bright prospects for the future if they continue to improve their leadership abilities.

Technical Communication Employee Levels

Technical communication groups can be organized into several levels or left with one large level. A typical arrangement is to have three levels, designated with something like TC 1, TC 2, and TC 3. Sometimes these levels are given more descriptive names, such as Junior TC, TC, and Senior TC, although terms such as junior and senior should probably be avoided. Hardly anyone wants to be called a "junior," and many of us are also sensitive to the appellation "senior." Managers of technical communication groups must also decide what the criteria will be for each of the levels. The criteria used for deciding what constitutes promotion from one level to the next are very important. What those criteria are will determine what employees work toward, so it is important for the criteria to match the missions and goals of the organization.

Several criteria are commonly used to determine how a technical communicator moves from one level to the next, including:

Performance. This describes how well the employee does the job as it relates to helping achieve the goals of the technical communication group and the overall organization. Performance is usually tied to the quality of the employee's output, but it can also include other factors such as timeliness, teamwork and collaboration skills, project management skills, etc.

Skills. This relates to what skills the employee demonstrates proficiency with. In a purely skill-based system, when an employee achieves prescribed skill sets, he/she is promoted regardless of seniority, education, or time in the job.

Seniority. This is based on one or both of two criteria: how long the person has been with the organization and/or how long the person has been in his/her current position. For example, you can require that a TC 1 be with your organization for at least two years before being eligible for promotion, or you can require that the person has been a TC 1 for at least two years, regardless of his/her overall time with the larger organization. While tenure with a company has its value, it can lead to a sense of entitlement. Some communicators grow considerably in five years; some grow very little. If at all possible, I recommend that seniority not be used as a criterion for technical communicator advancement. It can lead to significant resentment among younger employees whose skill sets, particularly in newer technologies, are superior to their "seniors."

Education. With the rapid rise in the number of colleges and universities offering degree programs in technical communication, some organizations are requiring such degrees for jobs as technical communicators. While I firmly believe in the value of such degrees, it is not a good idea to restrict technical communication positions to people with degrees in the field. I have worked with many excellent communicators who had degrees in biology, computer science, history, music, etc. While I also endorse technical communicators getting graduate degrees, I do not believe it is wise to tie promotion to the level of education, other than setting some minimum for employment in the first place, such as a bachelor's degree. This category might also include mandatory training and skill sets that employees must have before moving to another level, such as knowing how the organization's internal financial systems work.

Most systems combine aspects of each of the criteria to arrive at how people will move from one level to the next. Setting those criteria is important because it will affect how people work and what they aspire to. If timeliness is critical in your organization and your advancement criteria call for perfection, you have a mismatch between a key organizational goal and the criteria you are using to judge whether someone should be promoted. You may get employees missing deadlines just to take the extra time to ensure their documents are perfect. It is important, therefore, to match the criteria as carefully as possible to the real outcomes that you want. If the overall organizational goals change, you will probably need to change your job descriptions to accommodate those changes.

Some employees are more or less indifferent to moving to the next level; others are obsessed with it. If the system is too complex, a manager will have difficulty tracking who has accomplished what and negotiating with employees about which criteria they have met and which they have not.

Job Descriptions

A common method for listing the criteria is to write job descriptions for each of the levels you have. Job descriptions have major advantages and disadvantages. On the one hand, they are indispensable for informing employees about what is expected of them in their current positions and what they will have to do to achieve promotion to the next level. On the other hand, job descriptions can lead to stasis, with an organization moving too slowly to adapt new technologies and work processes simply because they are not in anyone's job description. Some organizations are eliminating job descriptions in the belief that in the information age all employees should work together to do whatever is necessary to accomplish the organizational goals and that job descriptions lock people into functional definitions when they may be capable of doing many different kinds of tasks and jobs.

Nonetheless, for organizations that have multiple levels, it is almost a necessity that job descriptions exist. Otherwise, there can be contention, including legal contention, about who gets promoted and why. The job descriptions should explain in some detail each criterion and how an employee meets it. It is better to have a few criteria explained in detail than numerous criteria vaguely explained. If teamwork is a criterion, explain how you will determine that someone's teamwork skills are good enough to make them eligible for the next level for that criterion.

Job descriptions may have some or all of the following parts:

- Job title
- Purpose of the position: Explains why the job exists and how it contributes to achieving overall organizational goals and objectives.
- Responsibilities, including percent of time spent on each: Includes main (4-6) responsibilities of the job, describing each briefly and giving the percentage of the time that an average person in the position will spend on each.

- Required knowledge: Lists the knowledge sets that someone in the position is expected to have. For communicators, this will include three types of knowledge: (1) the rhetorical and communication knowledge needed, (2) the tools and technology needed to design and develop the required outputs, and (3) the industry-specific technical knowledge needed (e.g., computers, telecommunications, pharmaceuticals, etc.).
- Experience required: Designates how much experience with the organization and/or in the same industry is required before one can move into the position.
- Education required: Designates the degree(s) required to move into the position. It is a good idea to include a category such as "or equivalent experience" to provide the flexibility to promote excellent performers who do not have the necessary academic credentials.
- Interactions: Explains the people with whom the job holder must work, emphasizing required collaborations.
- Primary challenges: Lists the three or four most difficult decisions or tasks that the job holder must succeed in.
- Responsibility and authority: Designates whether the position is supervisory or non-supervisory and/or how much authority for decisionmaking and budget commitment the job position entails.

The Job Description Template provides a guide for creating technical communication job descriptions. Normally, you would provide more information than space allows on the form, which has been compressed into one page.

Technical Communication Salaries

Prior to the development of personal computers, technical communicators earned considerably less than their peers in scientific, engineering, and programming positions. During the 1980s and 1990s, however, communicator salaries advanced well above the inflation rate as more and more organizations began to realize the importance of their relationships with customers, including communications to make their products and services more valuable. At least in some organizations communicators can earn as much or more as scientists and engineers.

If you have to develop salary ranges for communicators, the best source to study is the annual STC salary survey, which can be found on their Web site at http://www.stc.org. The survey shows ranges for both the U. S. and Canada, and it also includes averages for U. S. zip codes. This helps considerably in determining the appropriate salary ranges for your geographical area. Some of the local STC chapters also conduct salary surveys in their areas.

As is the case with promotion criteria, the method used to establish salary levels will affect how people work and what they aspire to. In general, it is a good idea to have salary ranges overlap, so that you have the flexibility to continue giving raises even when a promotion is not in order. Without overlaps, you face the problem that when employees reach the top salary in their pay grade, you

Technical Communication Management Worksheet

Job Description Template

1. Job Title:

2. Purpose of the Position (how does it help fulfill organizational goals):

3. Responsibilities: Describe four to six main responsibilities and the percent of time spent on each.

Responsibility	Percent of time
1	%
2	%
3	%
4	%
5	%

4. Required Knowledge:

5. Experience Required: 1-3 years   4-7 years   8-12 years   12+ years

6. Education Required: HS   Bachelor's   Master's   Ph.D.

7. Interactions: Explain each briefly
      Customers   Peers   Management   External

8. Primary Challenges:

Challenge	Successful Completion

9. Responsibility and Authority:

      Personnel: Supervisory   Non-Supervisory

      Budget: $0-500 $500-5000 $5000-50000   >$50000

TABLE 2.1	Sample Technical Communication Salary Ranges
Position Title	**Salary Range**
Technical Communicator I	$35,000–$50,000
Technical Communicator II	$40,000–$60,000
Technical Communicator III	$50,000–$80,000
Technical Communication Manager	$60,000–$90,000

either have to promote them or freeze their salaries indefinitely. Table 2.1 provides an example, roughly based on 2002 salary levels, which may change over the next few years.

Technical Communication Job Titles

What should we call those who perform technical communication jobs? Given one of the premises of this book, that technical communicators are developers, I lean toward a job title that includes the idea of development, such as the term used at IBM and other companies, "information developer." Almost everyone has abandoned the previously ubiquitous "technical writer" for two primary reasons: (1) practical—nearly all communicators now do much more than write; some are essentially programming as they develop Web sites and XML systems, and (2) political—writing is a skill that is less valued than developing or managing, and, in their desire to achieve equal pay and status communicators have learned not to refer to themselves as 'writers.'

One view of the future states that technical communicators will increasingly move toward managing gateways into large information databases, using their skills to create appropriate doors into single-sourced information storage repositories (Weiss, 2002). To the extent that we believe this to be true, we might consider using job titles that reflect storage and access activities for knowledge and information.

In general, job titles would ideally describe the things that people work on and what they do to those things. "Boilermaker" is clear enough. It is impossible to provide a single recommended set of job titles for activities as diverse as those performed by technical communicators. However, for both practical and political reasons, we should use terms that accurately describe what communicators work on and what they do to those things, using terminology that accurately depicts our growing responsibilities to create and manage things other than paper documents. Table 2.2 offers examples of terms associated with paper and terms associated with more diverse media.

For those in industries that are more specialized, try plugging in terms in the third column both for what communicators in your organization will work on and what they will do. Doing so can help you arrive at appropriate job titles in your specific case.

TABLE 2.2	**Technical Communication Job Titles**	
	X = paper	X = electronic, Web sites, user interfaces, etc.
TCs work on X	book, document, guide, documentation, manual, publication, report, training	communication, information, knowledge, product, proposal, Web
TCs do what to X	writer	communicator, developer, designer, facilitator, manager, specialist

The Hiring Process

Few activities are as important for a manager as hiring. The quality of the work your group creates depends largely on who the people in the group are. So does the quality of your work life as a manager. You will have to work closely with each of the people you hire, so it is important to follow a careful process when bringing in new employees. The hiring process has several crucial steps, which will be discussed in detail in the following sections:

1. Assessing Personnel Needs (Quantitative and Qualitative)
2. Recruiting
3. Evaluating Resumes
4. Interviewing
5. Evaluating Candidates
6. Making an Offer
7. Mentoring and Acculturation

Assessing Personnel Needs

Assessing hiring requirements requires that a manager consider two factors, the number of people needed (quantitative) and the type of people needed, recognizing such factors as experience, education, skills, etc. (qualitative).

Quantitative Assessment

With proper project management and planning techniques, a technical communication manager should know in advance when additional staffing will be required. Without this knowledge, the manager will encounter a crisis, with the current staff being overworked while the manager scrambles to get help on board as quickly as possible. It is important to become an active part of the organization's communication system regarding new projects, so that you can anticipate personnel needs earlier. As Hackos (1994) points out, you can do so through networking, meeting with marketing and other departments that initiate new projects, and getting information from technical communicators about new projects they are aware of.

If your organization does annual budgeting in which projects and their estimated headcounts are planned in advance, you can predict how many people you will need during the year. One way to do this is to use a table and/or a spreadsheet or a graph. See Table 2.3 and Figure 2.1.

In Table 2.3 and Figure 2.1, we can conveniently see the staffing needs for the entire year. If we have 10 communicators on staff now, we can consider using contractors until August. However, the increased needs for the last four months of the year indicate that we should consider hiring at least one or two permanent staff members, and three if we believe that the project demands will continue into the following years. That means we should begin getting approvals and preparing recruiting ads two to three months in advance, in May or June.

Qualitative Assessment

Before hiring, a manager should do a qualitative assessment of the group's needs. This assessment should cover what the current mix of the group is in terms of experience, education, and skills.

Experience

A manager must decide what experience range best fits into the group's needs. Upcoming work projections should help determine whether someone with many years of experience is needed or someone at the entry-level would be better. Many managers prefer to hire experienced people only. Such hires reduce the manager's workload because such employees do not require as much mentoring, training, and hand-holding as do entry-level hires. However, there are dangers to a group becoming too heavy with experienced people. First is cost. More experienced people cost more, and they are especially expensive if they are put on projects that do not require their levels of expertise. Not only is this wasteful, but it eventually will cause the experienced employees to become bored and to start looking for more challenging positions. Second, groups with only experienced people tend to move toward stasis. Entry-level employees, simply by asking why things

TABLE 2.3	Technical Communication Staffing Requirements— Year 200X											
Project	Jan	Feb	Mar	Apr	May	Jun	Jul	Aug	Sep	Oct	Nov	Dec
Alpha	2	2	2	2	2	2	3	3	3	3	3	3
Beta	3	3	3	2	2	2	3	3	3	3	3	3
Gamma	4	4	2	2	2	2	2	2	4	4	4	4
Delta	3	3	3	4	4	4	4	4	4	4	4	4
Total	12	12	10	10	10	10	12	12	14	14	14	14

Technical Communication Management Worksheet

Hiring Process Steps

1. Assess Personnel Needs — See Form.

2. Recruiting — Check methods to be used
   - ☐ Newspaper Ad
   - ☐ Our Web site
   - ☐ Employment Web site
   - ☐ STC Web site
   - ☐ College/University
   - ☐ Employee Referral
   - ☐ Agency/Search Firm

3. Evaluate Resumes

4. Interview Candidates — See Form.

5. Evaluate Candidates — Rank in order of preference.

   1. _____
   2. _____
   3. _____
   4. _____

6. Make an Offer

   Salary Calculation ___ $ _____

   Calculated by _____ Date _____

   Offer made on Date _____

   By _____ via

   ☐ Telephone   ☐ Letter   ☐ E-mail   ☐ Other

7. Provide Mentoring and Acculturation

   Report Date _____

   Mentor's Name _____

   Objective-Setting Meeting Date _____

   Other orientation activities _____

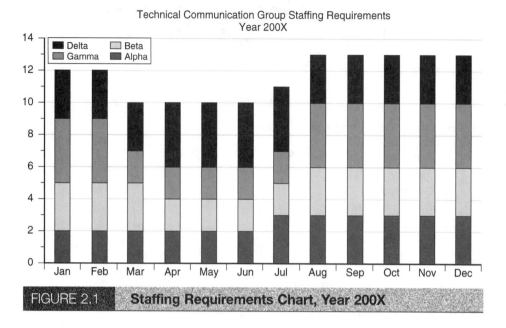

FIGURE 2.1   **Staffing Requirements Chart, Year 200X**

are done a certain way, can challenge people to think about possible improvements. If no one is asking such questions, it is too easy for everyone to become satisfied with the status quo. So it is a good idea for a manager to consider what mix of youth versus experience is needed in a group.

Education and Skill Levels

The same principle applies to education and skill levels. Depending on the nature of the industry, a manager may be able to hire communicators regardless of their educational background. However, in some types of industry and government work, a mix of people with bachelor's, master's, and even Ph.D. degrees may be required. So a manager should assess the current educational mix in the group and decide what level would work best for the demands on the group in the present and the future. If your assessment shows that you need special skills (graphics, translation, indexing) for brief periods of time, you should consider hiring contractors rather than full-time employees.

Tools versus Communication Skills

Likewise, immediate needs may point to someone who knows a particular tool very well. This is especially true if the hire is a replacement for someone who was a specialist in the tool. However, it may be a mistake to hire someone simply because of expertise in one or two tools. In the long term, it is far wiser to hire someone who knows how to analyze audience needs and how to design and deliver documentation products that meet those needs (Molisani 1999(2)). If you need a tool jockey for a quick fix, consider using a contractor rather than hiring someone full time simply to fix your short-term problem.

Fit with Existing Group(s)

Finally, a serious part of the assessment should concern how well the person will fit into the group(s) with whom he/she will work. This includes not only your group of communicators, but also the project team(s) and environment in which the candidate is likely to work. If you work with a group of tough, aggressive developers, you might not want an entry-level person. On the other hand, your group may need the questions and enthusiasm of such a person. Over many years of managing many groups, I found that the most effective ones were composed of a diverse mixture of age, education, background, gender, etc.

Obviously, you want to hire someone who will get along with the people in the technical communication group and who will be able to share the culture and the work environment of the group. This does not mean that such a person should share the same type of educational, social, or cultural background. Rather, it means that they should either have experience in, or show a clear willingness to learn, the collaborative practices of other group members. You do not want to introduce a control freak to an otherwise happily collaborating team. Nor do you want to bring in someone who will silently acquiesce to whatever is going on without offering any new ideas.

There is, however, a danger with working too hard to ensure that someone new will fit into the group. After awhile, you can end up with a group that thinks, acts, and even looks alike. You are in danger of entering into "groupthink," wherein you have a homogeneous, like-minded group that has not had its methods, assumptions, and processes challenged or improved for far too long. Such a group can settle into a kind of stasis, believing that they have arrived at the optimum set of methods and processes. Bringing in someone who does not quite fit might be the best thing you could do, especially if you suspect that you need to reconsider some of the methods and outcomes you are experiencing. Sometimes, bringing in someone from another industry or from a competitor can provide a technical communication group with insights into alternative and more effective methods and outcomes.

After doing both the qualitative and quantitative assessments, it is a good idea to prepare a set of objectives and characteristics for the position(s) you are about to post. This will help remind you, throughout the hiring process, as to what needs you uncovered during the assessment, so that you can stay focused on those needs during resume assessment, interviewing, and the remainder of the process.

Recruiting

Using the most effective recruiting sources is an important management skill. Arguably, finding the best employees possible is a manager's most important requirement because all other aspects of the job are strongly affected by how well a manager puts together a staff and how effectively that staff can work together and with others in the organization.

Hiring new employees takes considerable time and effort. It also costs money. Advertising, recruiting trips, interviewing time, agency fees, and other activities can add up to considerable expense for each new employee hired. Some technical firms

Technical Communication Management Worksheet

Personnel Needs Assessment

Date Needed _____

Quantitative

Number of Employees Needed _____

Duration Needed _____months

☐ Direct Employees — How Many? _____

☐ Contractors — How Many? _____

Qualitative

Experience
☐ Entry-Level
☐ 1-5 years
☐ 5-10 years
☐ 10+ years

Education
☐ BA/BS
☐ MA/MS
☐ Ph.D.

Skills
☐ Revision/Modification
☐ Editing
☐ Document Design & Development
☐ Documentation System Design
☐ Project Management
☐ Web site Design & Development
☐ Help System Development

Technical Comm. Tools
☐ Microsoft Word/Office
☐ Framemaker
☐ Interleaf
☐ RoboHelp
☐ ForeHelp
☐ Doc-to-Help
☐ Front Page
☐ Dreamweaver

Science/Technology
☐ Computer Hardware
☐ Computer Software
☐ Telecommunications
☐ Electronics
☐ _____

Objectives

Three Most Important Objectives for Hiring

1. _____

   _____

   _____

2. _____

   _____

   _____

3. _____

   _____

   _____

HR Contact Person_____

estimate that it can cost as much as $50,000 to hire a new employee, especially if we include the training time required to get the employee to productive activity after the hire. It is important, then, for managers to seek ways to recruit new employees that reduce the cost per hire as much as possible.

Where does one find good technical communication candidates, especially if there is a shortage of qualified people? There are many sources for recruiting technical communicators, and you should pursue all of them. The old, tried-and-true procedure of running an ad in your local newspaper's Sunday edition simply isn't enough.

In fact, there are numerous ways to reach technical communicators more effectively than traditional job advertisements. Those methods include referrals, STC chapter job banks, campus recruiting, networking, Web-based methods, and search firms.

Referrals

One of the most effective methods for recruiting is to get referrals from existing employees. Many firms offer compensation to employees who refer people who are eventually hired. Even if your firm does not, most employees are happy to try to help get acquaintances hired. Such acquaintances might include co-workers at former employers, friends from professional societies such as STC, alumni from employee's schools, personal friends, etc. The most effective method for eliciting referrals is to show your employees the result of your assessment, including the set of objectives you prepared. With that information, they can determine whether they know someone who might be appropriate for your opening. Once you get the acquaintance's name and contact number, you can either have your personnel department contact him/her, or do it yourself, depending on your organization's policies.

Referrals are a very inexpensive way to find new employees. Depending on your organization's policy, they may be free or may require a modest referral fee. Still, compared to the expense of using agencies or recruiters, this method is relatively low in cost.

STC Chapter Job Banks

Another low cost method is to search the Web site of the local chapter(s) of the STC job bank. Some chapters operate comprehensive job bank sites that include openings from hiring firms and resumes of candidates who are looking for positions. Because the job bank is free, this method costs very little. You can find the STC's chapter page at http://www.stc.org. From there, you can link to the closest chapter to your area and see if they sponsor an online job bank, or perhaps a job bank using voice mail, paper references, or other methods. With an on-line job bank, you can post your position to elicit applications. You can also search through the resumes listed on the site to see if anyone fits your requirements.

Because this method is so inexpensive, it makes sense to post all positions with the local STC job bank.

Campus Recruiting

Another relatively inexpensive method is to recruit on college campuses. You can do so in three ways: (1) by going to campus placement offices to interview potential candidates, or going to "career fairs" or "career days" when hiring firms are brought to campus to set up booths and talk to upcoming graduates about possible employment, (2) by developing an ongoing relationship with professors in the school's technical communication program, and (3) by working with the school's STC Student Chapter, if there is one.

The first method requires travel to the school, usually with brochures and information about your organization. It may require two or more people to staff a booth or to interview potential candidates. However, it can be a good way to recruit entry-level candidates, especially if the school has a technical communication program that offers degrees or certificates. If the school has a graduate program, you will often find upcoming graduates who already have some full-time experience.

It is a good idea to develop an ongoing relationship with the professors who teach in the school's technical communication program. You can often call or e-mail a professor and request the names of appropriate candidates for openings. This ongoing relationship can also help locate candidates for internships, summer jobs, and cooperative opportunities. Such temporary hiring situations can also help with ongoing recruiting because they provide a manager with an inexpensive method of evaluating current students while simultaneously getting some work accomplished without having to hire full-time staff or contractors.

If the school has a student STC chapter, you can contact its president and/or the professor who advises the chapter. They can often supply information about possible candidates and they may be able to advertise the position to all students on an internal mailing list.

Networking

Another inexpensive recruiting method is to network with other area technical communicators. You can do this at STC or other society meetings and at regional STC conferences. You can also do it through informal contacts with managers at other organizations, friends, etc.

If your area has a large group of technical communicators, networking can be an inexpensive and efficient recruiting method.

Web-based Recruiting

There are many ways to recruit technical communicators using the Internet. The best known method is to use one of the large employment databases, such as http://www.monster.com or http://www.anotherone.com. Using these services, you can both post your position and search through a large database of resumes using keywords to limit the search. The nature of these sites is constantly changing, but some of them do require a flat fee or a percentage of a new hire's starting salary.

Another method is to post job openings on your organization's home Web site. If your organization has such a site and it includes a "Jobs" section, you should be able to post your opening there. Using the set of objectives you developed will make it easy to prepare a list of job requirements for the online job posting.

You can also consider posting the opening to appropriate newsgroups and list-servs, although many of them discourage the bandwidth required by such messages.

Agencies and Recruiters

Some organizations have internal recruiters. If you work with internal recruiters, you should assist them with as much information as possible, as they are likely not to have extensive experience recruiting technical communicators. This includes carefully preparing lists of keywords for your openings because many recruiters use such keywords to search their own databases and to search on-line databases. While it is easy to provide them with keywords that describe various tool names, it is more difficult to find appropriate terms for the more important skill sets that technical communicators should have. That is another reason to prepare a set of objectives that spells out what you are looking for. You can provide the internal recruiters with a copy of the objectives. You may also need to provide the recruiters with some of the other recruiting resources described in this section in case they are not aware that such resources exist.

External agencies usually require a substantial fee for finding a new employee, on the average of 15 to30 percent of the new employee's first-year salary. Agencies work on two models: contingency and retained. In the contingency model, you provide the agency with your openings. They research and recruit candidates with matching skills and then send you the resumes, often with names and contact information removed. If you are interested in interviewing a candidate, the agency sets up the interview. If you hire the candidate, you then pay the agency a fee equaling 15 to 30 percent of the candidate's salary. In the retained model, you pay the agency a flat fee up front (usually equaling about 20 to 30 percent of the position's expected salary). The agency then carefully recruits candidates for your position and forwards some designated number of resumes, usually on the order of three to five. Once they have done so, they have fulfilled their obligation, even if you don't hire any of the candidates.

The contingency model has the advantage that it is free until you actually hire an agency's candidate. The big disadvantage is that contingency firms tend to work on a volume basis. They will often search a database they maintain using keywords they pick out from your requirements. Depending on how much expertise the recruiter has in your area, you may get numerous resumes of people who aren't actually qualified for the position. Their method is to get as many candidates' resumes as possible in front of you as quickly as they can, and hope that at least one of them works.

The retained model has the advantage that the firm usually spends considerably more time recruiting and qualifying candidates for your position, including interviewing them and learning their salary requirements at the outset. You will usually get some desirable candidates who want to work for your organization. The distinct disadvantage is that you have to pay up front for something that might not lead to a hire.

Another associated recruiting method is to engage contract employees from agencies and contracting firms. You can hire them for contractual periods with what is called a "right-to-hire" clause. This is a clause in the contract you set up with the contracting firm that allows you, after a designated period of time, to offer the contract employee full-time employment with your organization. Some contract firms require a fee (either flat or a percentage of the employee's first-year salary). Nonetheless, this can be a good way to assess the knowledge and skills of potential employees. This is a rather widespread practice among technical communication contracting firms, so you should check with the ones in your area to see if they have such "right to hire" policies.

Evaluating Resumes

Whether a resume is on paper or on-line, it can tell you much more about a candidate than its mere contents. A resume is a document. You are hiring someone to design and write documents. So when you are evaluating a resume you have right in front of you an example of the candidate's document preparation skills, including overall design methods, layout, and the use of fonts and white space to highlight main ideas and make scanning easy (Sopensky and Modrey 1994). You can also see if the candidate matched the resume for the particular type of position you have. A technical communicator's resume serves as an example of how well that person accommodates the needs of readers in a very specific rhetorical situation. If the resume is not effectively formatted to make information easy to find, how good are the chances that its creator will communicate effectively in other situations? The communicator has sent you a writing sample, which you can peruse without an interview. Use the resume to help assess which candidates look like better fits for your environment.

This is true even for files sent in ascii or rtf or other text-only formats. Even within the constraints of text-only, you can still look for headings, spacing, capitals, and other conventions that allow for easier reading of the resume. A resume is a particular kind of document sent to a particular kind of audience for a particular purpose. You can learn much about someone's technical communication skills from carefully analyzing the document he/she sends you.

In some organizations, employees in the human resources (HR) or personnel department screen resumes before they are sent on to hiring managers. If this is the case in your organization, you have two choices. First, you can try to get the policy overturned in the case of technical communicators because most personnel employees simply do not have enough experience or knowledge to assess communicator resumes appropriately. Second, you can work very closely with your personnel agent(s) to train them in how to assess the resumes for each specific search that you are doing.

If possible, a technical communication manager should look at all of the resumes received. A resume might arrive that describes an important skill using language that doesn't appear on the keyword list the personnel agent is using. Or a resume may arrive describing skills and experience that do not fit the job currently being advertised, but do fit a need you know you will have in the near future.

Before reviewing resumes, it is a good idea to look at the list of objectives you prepared for the position. You may be dazzled by a good resume, but if it does not fit the objectives you have for the position, you could end up hiring the wrong person. Also, reviewing the objectives will help you scan the resumes more quickly, looking for key words, types of experience, and other relevant qualifications.

Interviewing

Interviewing technical communication candidates requires that you employ some additional techniques to those considered standard for job interviews. An interview entails both assessing and recruiting the candidate. You can cover the recruiting piece, at least partially, by giving the candidate a good overall vision as to where his/her job will fit within the larger organizational goals. The assessment requires careful questioning about rhetorical knowledge and strategies, writing samples, and work team experiences.

Interview Process

1. **Include other managers and/or team members**. Many heads are better than one in assessing a candidate's suitability for the job and for working with a specific group of communicators.

2. **Don't hire yourself unless you are what you need**. Hiring managers often tend to look for people like themselves, which is not surprising. However, managers must guard against hiring people with experience levels, education, skill sets, etc. that are similar to their own.

3. **Avoid the first 30-second syndrome**. Much of the literature on interviewing points to a phenomenon wherein hiring managers tend to make decisions about people before they have even begun to interview them. Obviously, these decisions are based on appearance and general demeanor and not details about the candidate's actual experience and skills. Because it is impossible to assess the qualities of a technical communication candidate without a thorough interview, technical communication managers should be especially careful to avoid making quick judgments.

4. **Script the interview, especially its early parts**. This is important for several reasons. First, it ensures that you are interviewing all candidates in a similar fashion, so that your comparisons are based on an interview process that is at least somewhat repeated, rather than wholly individualistic for each candidate. Second, it ensures that you do not leave anything out when describing the position, your expectations, and how the job fits into the larger group, project, and organization. Third, it helps you and the candidate get through the first awkward minutes of an interview more easily. One effective script begins by talking about the overall organizational structure, your group's place in that structure, the place in your group that the candidate would occupy, and the importance of the contribution from that position to the organization's overall goals.

5. **Give the candidate an overall vision of the organization and what his/her role will be within it**. Remember that an interview is a dual process; you

are recruiting as well as assessing. Describing your organization's overall goals and how the candidate will contribute to them in the new position is a very effective recruiting technique. It also helps establish the context for determining if the candidate is interested in the position and if you are interested in the candidate.

6. **Questions to ask:**

   1. **Experience summary (including education, if appropriate).** Start with their most current position and work backward.

   2. **Why they left or want to leave their most recent position.** Red flags should go up if they complain about their boss or about management in general. Odds are they will do the same about you. A better answer is that they seek to learn about new technologies or that they seek greater responsibilities.

   3. **Detailed questions about writing samples, including rhetorical and design strategies.** Determine which part(s) of the sample they wrote. In interviewing dozens of technical communicators, I discovered that they often present writing samples they did not write. Ask point blank whether they were solely responsible for the entire document, or whether their role was to write a few sections, or simply to edit a few parts. Ask detailed questions about their choices for the overall document design, its organization, its page layout, font choices, etc. They should be able to explain why they made the choices they did to communicate most effectively with their audience. These questions will help you discern whether they have the skills to write and design for a specific audience or whether they have been in environments where they simply plugged information into templates without completely understanding what they were doing. The same sort of detailed questions should be asked about on-line samples, help systems, and Web sites.

   4. **Detailed questions about rhetoric, design, and audience focus.** To determine how much they know about these key areas, ask questions about how they have addressed them on previous projects.

   5. **Detailed questions about tools.** Many candidates will list the current, "standard" tools on a resume even if they have only a passing knowledge of how to use them. Ask a few details about some of the more esoteric uses of the tool to see if they truly know how it works.

   6. **"How would you....?"** Hypothetical questions about how they would design documents given specific situations. Hypothetical questions about certain audience/purpose situations can determine whether they understand what a good documentation process is and whether they will be able to respond to the new situations they will face with your organization.

   7. **Questions about teaming and collaboration experience.** Because technical communication work is increasingly collaborative and team based, ask some questions about their past projects and about how well they worked with others. Some communicators will proudly state that they

work totally independently and that they don't need anyone's help. Unless that is what you need, you can go through the rest of the interview quickly.

8. **Questions about experience with the science/technology with which your organization is involved and, specifically, with which their position would be involved.** If you are looking for someone who knows the technology your organization writes about, ask enough detailed questions to determine if the candidate understands it sufficiently well to work with your team and your subject matter experts.

9. **"How soon can you start?"** This is a trick question. You want them to show some enthusiasm for your position, but you also want to see if they would stiff their current employer without giving them two weeks' notice. If they will do that to their current employer, they will do it to you.

10. **Where do you want to be in one year? Five years? Ten years?** These questions help you determine if the candidate thinks ahead and has career aspirations related to technical communication or to something else. There are no right or wrong answers here, but the answers help you determine if the person is a good fit for your position and your organization.

11. **Salary expectations.** Many books and articles on how to interview tell candidates not to divulge their current salary and to put off salary discussions as long as possible. So do not be surprised if it is difficult to get a number from someone. A good answer is that the candidate wants your position and hopes that you will offer what you think their services are worth. The old method of determining someone's current salary and offering 10 or 15 percent more seems marginally ethical and ill-advised. If they will leave their current position for a 10 percent increase, they will leave yours too. If they will tell you their current salary, fine. If not, work with your personnel department to come up with a reasonable offer.

7. **Questions NOT to ask:**

   1. **Age, ethnicity, family, health.** Questions about these matters are against numerous laws, both federal and state. Even seemingly innocuous questions about spouse and children could be construed as being discriminatory, depending on the circumstances. It is better to stay focused on the position you have and their qualifications for it.

   2. **Proprietary issues about other organizations.** These questions can also be marginally legal. One technical communication interviewing trick is to ask if you can keep a copy of a proprietary writing sample they have shown you. If they allow you to, consider the possibility that they will someday give your organization's proprietary documents to others, too. The right answer for them to give you is no, that while they have permission to show the document as a sample they cannot leave it with you.

   3. **Personal questions.** Once again, it is better to stay focused on the position and the candidate's qualifications. If someone lists hobbies on

his/her resume, it is all right to discuss them briefly. Generally, however, you should be very cautious about discussing personal matters.

8. **Give them a chance to ask questions, and evaluate what they ask.** If all of their questions are about benefits, time off, and what they will get, raise a red flag. What you hope for are questions that indicate a genuine interest in doing the particular type of work you need. This also gives you a chance to see if they have done any research and prepared a list of questions, skills they are likely to need for your position.

9. **Ask an open-ended summary question**; e. g., "Why should we hire you?" You are hiring someone whose main task will be to communicate effectively. Asking an open-ended question allows you to see how well they can communicate about a subject they know and care about deeply.

10. **Conclude with the date by which you will follow up**. Make sure you follow up by that date, even if you have to contact the candidate on that date to say that you're still assessing possibilities (in which case you give them a new date). When you conclude an interview, you have an action item. You are responsible for communicating to the candidate, as quickly as possible, the results of the interview.

Evaluating Candidates

As soon as possible after the interview, compare the results to the list of hiring objectives you developed earlier. This will help prevent you from hiring the right person for the wrong job, even if that person's qualifications (for another job) were impressive. If you interview someone that impressive, you may want to create another position so that you can make an offer. However, it is important to make sure that you are qualifying candidates for the position you are hiring for and for the skill sets that you earlier decided you need.

Naturally, you have to take a number of variables into account here. If you have only enough funding for an entry-level person, you can't hire an experienced person who wants too much money. You have to look for the best fit with the list of criteria you developed earlier and relate it to the reaction(s) that you and others who conducted the interviews had to the candidates. Obviously, how well someone's personality seems to fit into your existing group and into the organization's overall culture is such a strong consideration that it might outweigh other criteria, such as experience in your industry or with the tools you use (Bryan 1994).

Once you have chosen the candidate to whom you wish to make an offer, you generally have to work through your organization's human resources group for approval to make the offer at a certain salary, although this varies greatly from one organization to another. Because HR departments often lack extensive experience in properly assessing technical communication resumes, they may need your help in coming up with an appropriate salary offer. It is a good idea to tell them how many years of experience candidates have and whether you consider their degrees to match your requirements. If they do not provide you with a salary figure that you think is competitive enough to hire the candidate, negotiate with them. If they do not have any other data about technical communication salaries, you can show them the results of the latest STC Salary Survey, available at the http://www.stc.org Web site. You can also check to see if STC chapters in your area have conducted salary surveys.

Some managers believe in squeezing every dollar they can when hiring someone, and that the process itself is something of a game. They make it a contest to see if they can get a commitment at the lowest possible salary. They believe that they should always offer someone a lower salary than what the candidate stated as a salary requirement, just to see if they will take it. You will get better employees who are happier and who stay longer (thereby costing you less in the long run) if you do precisely the opposite. Offer candidates what they asked for, or even a little more, and you will spend less of your organization's time and resources interviewing and hiring replacements in the future.

Making an Offer

If possible, you should make job offers directly to candidates, either in person or over the telephone. Hiring another person may not be all that significant for you, but for them it is usually a major decision, perhaps a turning point in their careers. You are still recruiting at this point, so the more personal attention you give a candidate the better. Candidates often make job decisions based, at least in part, on how well they think they will get along with the person who will be their supervisor (you, in this case). So the more personably you follow through with their recruitment and hiring, the more likely they are to accept your position.

One exception to this policy is that it may be preferable to have them discuss benefits with someone in HR. With benefits packages often having complex mixtures of possibilities and of eligibility dates for a new hire, you might promise a benefit that you can't deliver later. A candidate is going to take everything you promise verbally as a contractual commitment from your organization, so it is better to have someone who knows the benefit system well to explain it to candidates and new hires.

Your offer should include the following:

- The job title (and level, if relevant)
- The start date
- The starting salary
- Any bonuses or commissions for which they would be eligible, and when
- The location where they will work
- The project they will work on, especially if you discussed it during the interview and they showed enthusiasm
- The date by which you need a decision.

It is a good idea to refer the candidates, at this point, to HR for benefits-related issues, and to give assurances that you will be happy to answer any work-related questions now or during the time they are making a decision.

Once the candidate accepts, it is your responsibility to arrange, through HR, for him/her to report on the start date and to sign all of the necessary forms (taxes, insurance, etc.) with HR on that date. It is also your responsibility to arrange for a desk, a telephone, a computer, office supplies, etc. The more professionally all of those essentials are handled, the better the first impression your organization will make.

Mentoring and Acculturation

The final step in hiring new employees is to begin the process of introducing them to your organization's culture and way of working. This process can begin even before they report for the first day of work. You can send them organization or project style guides and similar information to look over before they report to work. This could include project engineering specifications, requirements documents, and documentation plans. Note, however, that proprietary issues may preclude sending project information to them until they have actually signed all of the employment forms. You should be able to find enough non-proprietary information to get them started in learning about how you work.

Once the employee arrives, it is important, as Bryan (1994) points out, to state explicitly your expectations and your standards. It is a good idea to have an objective-setting session with new employees when they first arrive. Giving them a written set of objectives tells them exactly what you expect of them and what they should be working toward in their first months on the job. Further, you should give them copies of any standards, style guides, or principles for which you will hold them accountable. While much of the work of introducing a new employee can be given over to an official mentor or to fellow employees, it is important for a manager to take the lead in working on performance expectations.

One of the best ways to ease the transition of a new employee is to use a mentoring system (Smith 1993). Managers often do not have time to mentor someone adequately. They are often not the best choice to do so anyway because they may be far enough away from the experience of being a new employee that they have forgotten some of the things that need to be conveyed. In a technical communication group, the logical mentor is a more senior person who is working on the project to which the new employee will be assigned. That person will not only help to introduce the employee to the organization's culture and working methods, but can also explain the specifics of the project's technology, its documentation set, and the tools being used to develop the documents.

Working a new employee into the development process necessarily involves a trade-off between lost productivity and adequate mentoring and coaching. While it would be nice to have the employee immediately begin to do productive work, it is hardly fair to expect that from day one. Even experienced workers who have come from other organizations will need some adjustment to a new organization and especially to its work processes. A mentor can help greatly to determine when the new employee is ready to begin doing project work, often by giving the person short project assignments to complete and guiding them through the process. How a mentor and manager revise the employee's first writing efforts will help to inform the employee about what is expected (Katz 1999). Ultimately, the manager must decide when to press the new employee into a full work schedule. The credibility of the manager and the manager's organization will be well-served by not doing so too quickly.

It is important to make the learning process a two-way street. Indeed, the manager and mentor want to introduce the new employee to the organization's culture. However, they also want to learn from the new employee's questions and problems. A new employee's questioning of steps in your development process might help you see that you are doing something unnecessary and that you've become blind to its redundancy "because we've always done it that way." If the new employees have experience working at other organizations, their insights into how similar processes were conducted elsewhere might contribute to improved processes in your group's work. In other words, while it is important to acculturate a new employee, it is also important to improve your culture and work methods by learning from the employee's questions and knowledge before completely inculcating your organization's own methods and values.

Hiring Contractors and Consultants

If your original assessment shows that you need to hire contractors or consultants, you should follow most of the same steps as you do when hiring a full-time employee. Of course, the recruiting method is different because you will likely go to agencies rather than seeking other recruiting venues. And you are less likely to follow each of the steps as intensely as you would when hiring an employee. Still, contractors can play a very important role on some projects in some organizations. It is important, therefore, to apply much of the same planning and care in hiring them as you do in hiring employees. This includes care not only in the assessment and hiring process, but also in ensuring that the consultant is trained and acculturated well enough to work effectively in your organization (Thuss 1982). Contractors can sometimes bring work habits and assumptions that are based on past experience but that may not be appropriate for your environment. Also, you want the contractor to make the most positive contribution possible to your work. While some organizations refuse to spend any money or time training contractors, it is worthwhile to at least spend some time training the contractor on your own work processes, style conventions, tools, and other aspects of the organization (Armbruster 1986).

Contractors are usually paid at an hourly rate that is considerably higher than the hourly salaries you pay for a full-time employee. If the average salary of a technical communicator is around $50,000 per year, that translates to roughly $25 per hour. In many areas, however, you will have to pay $35-$50 per hour for contractors. The rate is higher because the contractors must recoup enough from what they charge to cover their own costs for taxes and benefits. When comparing employee and contractor rates, you should use the loaded rate (see Chapter 3) for your employees versus the contractor's rates. For example, if your loaded rate for an employee is the salary of $50,000 times 250%, you have an annual cost of $125,000, which equals an hourly rate of roughly $62.50. A contractor at $50 per hour is actually cheaper, depending on whether or not your organization adds loadings to contractors. You should be able to find out from your supervisor or your finance department whether your department pays loadings for contractors and how much those loadings are. If you then add those loadings to the contractor's rates, you can get an accurate picture of the cost comparison between full-time employees and contractors.

One important consideration when hiring contractors is the impact that it has on your employees. You can suffer morale problems if contractors tell employees what their hourly rates are (Alexander 1998). Many employees do not understand how loadings work and how much it actually costs to have them occupy a seat. All they know is that the contractor is making $50 per hour and they are making half of that. If you are about to hire contractors, it is a good idea to explain in a group meeting how loaded rates work and that, to arrive at a fair comparison, the cost of all the employee's benefits (taxes, health insurance, 401k, pensions, bonuses, etc.) must be added to the employee's rate . You will also need to explain that contractors are paid a premium for the risk they take of not having full-time employment.

Another important consideration when hiring contractors is the law as it relates to co-employment. Many technology and scientific firms hire contractors and keep them on staff for years without ever converting them to employee status. The contractors may be well compensated but they do not receive the organization's benefits nor the same legal protections that full-time employees do. Some of these long-term contractors have successfully sued the organizations on the basis that, after years working at the same jobs, they were essentially employees who were being denied the legal and compensatory advantages enjoyed by full-time employees. In some cases, damages and "back" benefits have been awarded. In general, it is not a good idea to keep a contractor on the payroll for more than a year. If you want to keep contractors on your staff for longer than that, you should hire them as employees. If you do not, you may face legal problems.

Despite the possible problems, contractors can provide many benefits. They can fill in for brief periods when work peaks are occurring. They can offer needed relief to stressed employees who are pushing to meet tight deadlines. They can provide particular skills that are needed only for a brief period. And they can bring new perspectives and ideas about work processes.

Training and Development

In the information age, the means by which we develop and transmit information are evolving rapidly. The dynamic nature of our industry mandates that technical communicators constantly learn new technologies for developing and delivering their information products. Moreover, the science and technology about which they write is similarly changing constantly, further increasing the necessity for frequent updating of knowledge and skill sets. Another contributor to the need for training is the evolution of many organizations toward non-tangible, knowledge-based assets rather than physical products.

In such an environment, a technical communication manager must ensure that employees receive ample funding and time to keep up with the changes going on both in the technologies we use and in those we write about (Scott 1996). The old idea that employees can be sent to one or two training workshops per year to keep their skills sharp is simply not enough. A manager must work with each employee to set up a training plan and to help the employee get enough funding and enough time to continue learning. Further, the manager should make it a part of the technical communication culture that everyone is learning all of the time, including promoting the exchange of ideas and knowledge among employees.

Professional Development for Technical Communicators

The various knowledge and skill sets required of technical communicators are so broad as to defy simple classification. The specific types of development most needed depend on multiple variables, such as the experience (or lack thereof) of a communicator, the length of time worked in a given industry or discipline, the length of time worked for a particular entity within the given industry or discipline, the specific social, political, and economic forces within the entity, and even more specific considerations related to a particular boss, work group, and/or project. In addition to weighing all of these variables, a technical communicator and his/her manager must also assess the amount of time and funding available to work on developing additional skills.

How does one work through all of these considerations to arrive at a development plan? This section breaks down development needs into five main categories. Within each of these categories are requirements for general development requirements and organization-specific requirements, which lead to ten main kinds of development that must be weighed against the time and money available. The section then provides a list of various development methods, ranked according to average costs. By building a table with a particular technical communicator's needs in each of the ten categories, and then weighing cost/time considerations against each type, it is possible to create for each communicator a development plan that optimizes development opportunities within budget and project constraints.

The five main categories include:

- General management skills (time management, budgeting, project management, personnel management) .
- Job-specific management skills (technical communication management, documentation storage and retrieval management, production management, etc.).
- General technical skills (trends in technical communication, on-line and Web-based technologies, distance learning and delivery, CD-ROM and DVD production and delivery, etc.).
- Specific technical skills (Framemaker for document development, Robo-Help for online help development, html, tools for developing Web pages and sites).
- Industry- or discipline-specific technical skills (related to the science or technology of the industry in which one works and the particular products or projects on which one works).

Within each of these categories there are two potential subdivisions: (1) general skills important in any industry and (2) specific skills related to a particular industry, technology, or project.

Table 2.4 includes the five main categories and the distinctions between general and organizational development needs. It also includes examples of each type of development required.

Developing a Training Plan

Ensuring that you meet each employee's requirements for training and professional development requires that you develop a training plan for each (Houser 1998; Hopwood 1999). Professional development is simply too crucial to be left to chance.

With the training needs identified in the table above, it is now possible to work with each employee (including oneself) to complete the process of developing a training plan using the following steps:

1. Assess training/development needs
2. Survey training methods available for each need
3. Develop an appropriate training plan

TABLE 2.4	Technical Communication Training Needs	
	General Development Needs	**Organization-Specific Needs**
General Management Skills	Time management, finance/budgeting, project management, personnel management	ABC's Budget Process, ABC's Quality Management System, ABC's Personnel Policies
Tech Comm-Specific Management Skills	TC management, documentation storage and retrieval management, production management	ABC's Document Storage System, ABC's Production System, ABC's Product Development and Delivery System
General Technical Skills	On-line and Web-based technologies, distance learning and delivery, CD-ROM/DVD production and delivery	ABC's Computer System, ABC's Product and Delivery System
TC-Specific Technical Skills	Publication development methods and software, on-line system development methods and software, html and Web-based development and delivery	Framemaker, RoboHelp, Dreamweaver, MS Word, Front Page
Industry- or Discipline-Specific Skills	Technology involved in general industry (i.e., telecommunications)	Company-specific technology within the overall industry (i.e., fiber optic cable, wireless communications)

(1) Assess Training/Development Needs

The first step in developing a training plan for each employee is to assess where the employee stands with each of the types of training/skills described in Table 2.4. Some organizations use a training/development schema that includes both specific training requirements (e.g., the internal training course in ABC's time reporting system) and general requirements, which can be fulfilled by more generally available courses, seminars, or books (e.g., introduction to project management). For example, see Table 2.5.

For each of the Skill types, the table includes the required courses/solutions. In most organizations, there would be more entries than in this example.

TABLE 2.5	Example of an Organizational Training Plan			
	Technical Communicator I	**Technical Communicator II**	**Technical Communicator III**	**Technical Communication Manager**
General Management Skills	The ABC time reporting system	Covey time management course	Intro to project management Understanding ABC's budget system	Advanced project management Advanced time management ABC's financial management system
Tech Comm-Specific Management Skills	Using ABC's document management system	Editing technical documents	Project estimation Information architecture	Advanced information architecture Advanced document management Advanced project estimation
General Technical Skills	Using the ABC computer network	Distance learning and delivery	Multimedia development and delivery On-line and Web-based technologies	Advanced network architecture Advanced on-line and Web-based technologies
TC-Specific Technical Skills	The ABC document development process Introduction to Framemaker	Advanced Framemaker RoboHelp	Special Framemaker skills Dreamweaver	Publication development methods and software On-line system development methods and software
Industry- or Discipline-Specific Skills	Introduction to our industry's technology	Advanced industry technology	More advanced industry technology	More advanced industry technology

If your organization does not have such a system, then you will need to develop an alternative method of assessing each employee's needs. This will need to be done even with the organizational listing because the needs, especially of experienced communicators, vary so greatly. Completing a Training Needs Assessment with each employee requires the following steps:

1. Distribute the Needs Assessment Form to all employees and have each enter what they believe their training needs to be.

2. Complete an assessment for each employee, taking into account the employee's last performance review (including areas for improvement) and any organizational mandates.

3. Meet with each employee to review the forms, reconcile them to the extent possible, and arrive at final agreement about which areas should receive priority.

After completing the assessment, it should then be possible to find the specific training means to be used in fulfilling each of the employee's needs. This could include any of the methods discussed in the skills development methods below.

(2) Survey Training Methods Available

The technical communication field offers many opportunities to develop employee skills. Because a requisite skill for technical communicators is the ability to research and find information, many communicators discover sources for continual learning on their own. The following sections describe some of these potential sources and provide some examples of places to look first, with the caveat that such sources tend to come and go quickly. For more detail, see Chapter 9.

Books

As the technical communication field grows, more and more books are published for the discipline. The advantages of having employees learn through books include lower costs, the ability to read incrementally during "slow" times rather than dedicating several days to a training class, portability, durability, and transferability (i.e., several people can read the same book). Further, employees can read books on their own time rather than during work hours. One way to show you are serious about a new method, piece of software, or technological development is to buy all of your employees a book on the subject. The time and expense of making the purchase and distributing the books demonstrates that you are serious.

Several publishers frequently issue books related to technical communication, most notably:

Ablex
   (affiliated with JAI Press)   http://www.hcirn.com/res/publish/ablex.html
Allyn & Bacon
   (affiliated with Longman)   http://www.ablongman.com/
Baywood   http://www.baywood.com/
John Wiley & Sons   http://www.wiley.com/

Society for Technical
   Communication   http://www.stc.org/

Classroom Training

Classroom training can be especially effective for cases where exposure to a concept or skill is needed quickly. New managers, for example, will benefit from taking training courses in basic management skills as soon after promotion as possible. In such courses they can quickly learn about those areas where they have not had prior training or experience, most often in personnel management and in budget/financial management. Training courses can also be very effective for learning about a new technology that a person or group is going to have to document, or about new techniques or software products that the group is adopting for use in preparing documentation. You can look for several different kinds of training, including in-house training, in-house training with an external trainer, external training, distance learning, and computer-based training.

In-house Training

In-house training has the advantage of being close, requiring minimal travel expense, ensuring that concepts are taught "our" way, and reducing disruption of the normal work schedule. In some cases, internal courses are the only source for a particular type of information (e.g., The Aardvark Budgeting System) and so will be the only place to get the training.

    In-house training with an external trainer provides most of the benefits of general in-house training, except that an outside training expert leads the sessions. This type of training is necessary when no one in-house has the knowledge set to develop a particular type of training or when your organization is using a product for the first time (a software development product such as Framemaker, for example). Many training consultants are willing to come to your site to train, even if their literature does not mention it, so do not hesitate to contact someone with the expertise you need and ask if they will do a training session at your site. Many trainers are also willing to customize their training so that it fits your needs as closely as possible.

External Training

External training has the advantage that the trainees are off-site and away from all of the distractions that occur at their work locations. They can become more immersed in the subject. It is also possible that off-site training is required by the nature of the subject (e.g., a flight simulation laboratory). The obvious disadvantages involve travel and lodging costs, "lost" time from work, and the cost of the training itself. See Chapter 9 for a list of organizations offering training designed for technical communicators.

Distance Learning

Distance learning provides numerous benefits. Again, the most important is that employees can complete such training while still at work or, potentially, while they are at home. Many universities offer courses in technical communication via distance learning, with more being added every semester. You can also find training for some software development tools via distance learning. Moreover, there are distance courses that teach specific courses or skills that people in your group need (including management skills for you). See Chapter 9 for a list of organizations offering distance learning opportunities.

University Courses

Courses at nearby colleges or universities are often valuable and relatively inexpensive. Some universities offer seminars and training sessions in computers and specific software programs for very reasonable rates. Search on the university's Web site to see what is available. Another option to consider is university courses taught by way of distance learning. Search the Web using the course subject you are interested in.

Computer-based Tutorials (CBT)

Self-paced training using computer-based tutorials can also be done on site or at home, and they offer the obvious advantage that they can be done incrementally, at one's own pace, without any dedicated times required. They usually allow sections to be repeated when someone needs to review a certain part or needs a refresher before continuing to a new section. Again, CBTs are available for many of the software tools we use, and they are often provided with the products.

Magazine and Journals

Magazine and journal subscriptions provide ongoing information and "training" related to our profession and the technologies we must write about. One of the best ways to learn a new field is to subscribe to its journals. Even if one is not going to develop the depth of knowledge that, say, a mechanical engineer has, subscribing to "Piping Monthly" will teach the technical communication manager about the concepts of the field, the jargon used, and new developments. The manager can also learn about competitive products and how they are advertised and documented. After serving some time in the field, the communication manager may find that some of those competitors are potential employers or clients for contract work.

For more information, see the list of journals in Chapter 9.

Memberships in Professional Organizations

The Society for Technical Communication is the "standard" organization for technical communicators to join. There are, however, many other organizations in our field and related fields to which technical communicators belong. Organizational membership provides ongoing training opportunities to members, usually through one or more journals, newsletters, books and other publications, local chapters, national and regional conferences, special interest groups, Web sites, mailing lists, and organization-sponsored training sessions delivered via traditional classrooms and via distance learning.

Indeed, the benefits of belonging to one or more professional organizations are so great that all technical communicators should do so. And the wise manager will pay for them to do so. This is a sore subject in many organizations. Some simply enforce a policy that no professional dues will be paid, but their parsimony is ill advised. Spending $100 or so per employee to pay professional dues sends several very strong and desirable messages. It says that management considers its employees important, that the organization is meant to be a "learning organization" or a "knowledge organization" (to use current buzz words), that it encourages employees to participate actively in their professions, and that it encourages employees to continue learning constantly. Refusing to pay employee dues sends the converse message.

For more information, see the list of professional organizations in Chapter 9.

(3) Develop the Training Plan

Using the Training Plan Form, the manager and the employee can fill in for each category the training requirement, the means that will be used to fulfill it, and the date by which it should be done. This exercise is very valuable for giving employees a picture as to what they will be learning and when. It also helps with retaining good employees because people are more likely to stay in a position where they know that their professional growth will be encouraged and aided. It helps the manager ensure that employees are getting training in the areas where they most need it, and it helps provide for the maximum development possible for the funding available.

Job and Project Assignments

Technical communication managers must make job and project assignments while weighing several variables:

- **The project**. How well will the people assigned to a particular project succeed? Will they contribute toward achieving the project goals and, in a larger sense, toward the organization's goals? Do they have the knowledge and skills necessary to do quality work?
- **The technical communicators**. How well does the project, including its technology and its documentation requirements, match the interests and aptitudes of the individual technical communicators assigned to it? Will they be engaged and challenged, or will they be bored and feel stifled? Is the science or technology involved something they care about or something to which they are indifferent?
- **The manager**. Will this assignment mean more or less work for the manager? Assigning a person with incomplete project management skills to an extremely complex project almost guarantees that more of the manager's time will be required. Do you have that much time?
- **The social and political considerations**. Will the person you assign succeed in working with the other groups and individuals that the project will require? Are you about to send a lamb to the slaughter? Or a bull into a china shop? Will the project team readily accept the entry-level person you propose to send? Is this a highly visible project or one that is less noticed?

Technical Communication Management Worksheet

Training Plan

Name _____   Date _____

	Job Title 1	*Job Title 2*	*Job Title 3*	*Job Title 4*
General Management Skills				
Tech Comm-Specific Management Skills				
General Technical Skills				
TC-Specific Technical Skills				
Industry- or Discipline- Specific Skills				

- **The opportunities**. Is this an innovative, challenging project where those involved are likely to grow professionally and to be noticed widely? Or is it a maintenance project on old technology where the people involved are unlikely to grow or to get much credit or notice, even if they do excellent work?

Project assignments can have a significant effect on an employee's opportunities for growth, performance rankings, promotion, and morale. A technical communication manager must balance the needs of the overall organization with the preferences and interests of employees when assigning people to projects. This can require decisionmaking involving a very complex set of often competing variables.

For example, if you know that the young, innovative person you hired last year is chafing under the boredom of the maintenance project you have assigned him/her, do you assign him/her to the new, high-tech, highly visible project that is coming up, or do you assign a more experienced person? Assigning the veteran could be safer, resulting in better political interactions but a less inspired document. It could also result in a resignation from the innovator, who finally gives up hoping for a more challenging assignment.

Should you always assign your best performer to your best project? While this seems intuitively to be the best policy, there are problems with it. First, it sets up a self-fulfilling hierarchy that tends to make it very difficult for less well-rated employees to succeed, even if they have grown and learned considerably since the last time they were rated. Second, it contributes to the "halo effect," such that once an employee comes to be highly regarded, he/she remains highly regarded even if the quality of his/her work no longer merits it. Being assigned to the most exciting, innovative, and/or visible project allows an employee to continue to shine, even if someone else might have done a better job. And it prevents those others who might do better from ever getting the opportunity. Third, it contributes to the perception that managers "play favorites," that they have favored employees who get the plums, and that others are condemned to work on the more mundane, less challenging projects.

A manager can employ some of the leadership skills discussed at the beginning of Chapter 2 to make job and project assignments work better. In particular, giving employees a vision and a set of objectives for each assignment can help the employee see the importance of the project. Some managers go so far as to designate what the employee must do on the project to receive a rating of "excellent" or "greatly exceeds expectations," of "good," of "average," etc. If you cannot conceive of anything an employee can do on a project for his/her work to be considered excellent, be aware that you are putting that employee at what may well be an unfair disadvantage, and that the employee will figure that out fairly quickly.

Even the most mundane project can afford opportunities for superb work, especially if the manager provides adequate vision and support for the employee working on it. Peters (1999) gives the example of a person who was assigned a project to clean up the company's warehouse and who ended up restructuring the entire workflow of the of the company's production facilities. Technical communication maintenance assignments can still provide opportunities to improve overall document structure, remove extraneous information, improve graphics, etc. If the budget and schedule limitations prevent anything but the barest updates to be made to documents, the manager must deal with the fact that the employee assigned to that project must either enjoy such work (and there are such people) or be rotated off onto something more interesting after a reasonable interval.

Technical communication managers must remember that project assignments greatly influence employees' morale, opportunities for growth, chances for promotion, and day-to-day job satisfaction. Getting the organization's work done well is important, but so is keeping satisfied, productive employees on staff. Careful job and project assignments can help do so.

Performance Evaluation

Performance reviews are perhaps the most difficult aspect of management, especially for the new manager. It is difficult enough for veteran managers to sit in judgment of their staff, but often painfully stressful for newly promoted managers to do so when they may be evaluating people who a few months ago were peers and, conceivably, even respected senior employees. However, it is a requirement of the job. Further, it need not be so traumatic if one follows some simple but important concepts. This section will discuss those concepts and then provide a sample evaluation form.

Evaluating technical communicators can be especially difficult because they often work on extended projects that can last for months, and, in some cases, even years. Assessing a communicator's progress mid-project makes the performance evaluation even more trying but also all the more needed. Because communicators often work for extended periods without much feedback, a technical communication manager should seriously consider implementing a formal performance evaluation system, even if the larger organization does not have one. Doing so helps to provide communicators with more frequent feedback than they would otherwise receive, which can help keep morale and productivity high. While implementing such a system may seem to add to a manager's workload, in the long run it should reduce it. Employees appreciate receiving feedback, even if it is not part of some larger organizational system.

Following are the basic concepts that should help managers do better performance evaluations.

Evaluate Against a Set of Written Objectives

The worst performance evaluations systems are those where managers unilaterally write performance reviews at the end of the year without any written set of criteria. This allows (perhaps even encourages) managers to be arbitrary, to play favorites, and to make things up as they go. Such evaluations give the whole concept of employee evaluation a bad name. Performance evaluation should not be a once-a-year phenomenon. Rather, it should occur throughout the year, with technical communication employees receiving feedback during each project and especially getting a mini-performance review at the end of each project. Such feedback is far more valuable when it comes just after a project concludes rather than several months later. Further, performance evaluation should be based on some mutually understood, written criteria.

At the beginning of the year, meet with each employee to draw up a set of objectives for the year. The employee should bring to that meeting a set of objectives that he/she wants to pursue, and the manager should likewise have prepared a set of objectives that the employee should pursue. The first place to look for each employee's new objectives is on their most recent performance evaluations. Those areas that you identified as needing improvement should each have an objective associated with it for the ensuing year. A second source that managers should consult to derive objectives is the set of goals and objectives developed for the entire organization. In fact, every objective that goes on an employee's performance review sheet should be related to the overall organizational objective that it helps to meet. This is especially important for communicators because it helps us to avoid the temptation to become a separate group of writers whose goals and objectives are personal and not related to the organization's goals. Because communicators often do work that is vastly different from most of their peers at a technological or scientific organization, it is tempting for us to judge technical communicator's work against our own sets of documentation quality criteria rather than against larger organizational goals. However, we should avoid that temptation and ensure that each goal an employee has somehow ties in with the larger, organizational goals.

The first step, then, in performance review is to meet with the employee, go over the objectives that you have each developed, reconcile the differences between them, and then agree on a formal set of objectives for the year. You can use The Annual Performance Objectives Worksheet as an example.

Make the Performance Review a Process Rather Than an Event

In their famous book, *The One-Minute Manager*, which still hits the best-seller lists occasionally, Blanchard and Johnson (1982) emphasize the importance of providing ongoing positive and negative feedback to employees, soon after the event that warranted such feedback occurs. While it is probably a good idea to give feedback in sessions that are longer than one minute, their point is still a good one. Performance feedback should be an ongoing, repeated process throughout the year rather than a single, awkward meeting that occurs at the end of the year. Ideally, employees should have a good sense of what is going to be in their performance reviews before they occur because they should have received feedback throughout the year (Dalla Santa, 1990).

Technical communication managers should meet formally with each employee for interim reviews at the end of each project. If that is not possible, they should meet quarterly or, at the very least, semi-annually. These meetings can be more informal and relaxed, and can take some of the tension out of the end-of-the-year review. However, if you are giving someone negative feedback, it is a good idea to make the interim sessions more formal. (See Performance Problems.) Interim reviews are particularly important for new employees, who will be eager to know how well they are doing at meeting your expectations.

Technical Communication Management Worksheet

Annual Performance Objectives

Year _____   Name _____   Date _____

Objective	Measure	Date	Reason	Status

Use this form to develop objectives with each employee at the beginning of the year.

Under Objective, put each objective that you want the employee to complete.

Under Measure, include the specific measure you will use to determine if the objective has been met. This should be quantifiable, if at all possible.

Under Date, put the date by which the objective should be completed. Many people assume that you can simply put December 31st for all annual objectives. However, this allows procrastinators to put things off until the end of the year, when they must scramble to get things done. It is much more effective to put some objectives in each quarter so that you can assess the employee's progress throughout the year.

Under Reason, put the reason why the employee must fulfill this objective. You might refer to an organizational goal or objective that the employee objective will help fulfill. You can also refer to the fact that it was identified as an area of improvement on the last performance review. Or, it could simply be something that the employee wants to do to broaden skills/knowledge.

Under Status, you can use the form to hold interim reviews and to track whether the employee is behind or ahead of schedule in completing objectives.

Another good way to make evaluation a process is to have employees participate in doing their own evaluations. This may sound strange, but having employees write their own performance reviews is an excellent way to begin the process early and to have employees accept whatever rating or ranking they receive on the final evaluation. It also helps prevent the notion that you have evaluated them against some arbitrary set of criteria. If you have the employee write the review and submit it to you electronically about a month before it is due, you have time to make changes (removing the superlative adjectives strewn throughout and adding some areas for improvement). You can then meet with the employee, go over the changes, negotiate the final version, and have the employee make the final changes and submit it to you prior to the performance review meeting. You, of course, must reserve the right to add any final changes even if the employee does not agree with them. Such a process makes the review meeting much less traumatic for both of you and it leads to employees believing that they at least got a fair shake, even if they may not agree with exactly how you ultimately rated them.

Base Performance Reviews on Performance, Not on Personality or Traits

One of the most common mistakes managers make is to base performance reviews on personality characteristics that are not related to performance. A performance review should evaluate performance against a known set of criteria for a specified period of time. In other words, you should base your evaluation on the objectives you agreed to at the beginning of the year (with modifications during the year to reflect changing assignments and organizational goals). Further, you should evaluate the employee's performance for only the period since the last formal review. It is not fair to say that someone did an excellent job on everything they wrote this year but that their performance last year was lacking so that their overall rating is average. If their work for this year was excellent, then their rating has to be excellent.

It is unwise to evaluate employees based on their personalities or on how they behave, except to the extent that those things affect their performance. If it is important for them to work amicably with other team members in the communication group and other groups in your organization, then doing so should be explicitly stated as one of their objectives.

Further, performance reviews should evaluate *performance*, not other factors. A brusque personality is not a liability unless it affects performance. An employee with such a personality should be told how it affects performance, not that it bothers you or some of the other workers (even if it does).

Make the Review Session a Pleasant Experience Rather Than a Trauma

One of the most important things a manager does each year is to have the formal performance review with each employee. How the manager conducts such reviews can affect employee morale, productivity, and turnover. And studies suggest that employees perceive that many managers conduct reviews poorly, with one survey showing that 80% of employees perceived that the review process is invalid and unreliable (McDowell 1994). Review meetings, then, should be carefully planned and carried out, using the following guidelines.

1. You should schedule ample time for each meeting. If a manager schedules fifteen-minute review sessions, the message to employees is clear. Either the manager is a coward and is afraid to discuss performance at length, or the manager does not really believe in the process. Scheduling an hour sends an entirely different message.

2. Consider having the review session somewhere other than in your office, which, from the employee's view, is the most intimidating place possible. You might try a conference room, the cafeteria (if you can avoid interruptions), or even somewhere off-site. At such locations you are more likely to have a relaxed, frank discussion about the employee's performance and about issues that affect it (Roff 1988).

3. Plan for the review session and prepare a "script" for it. The goal for a review session should be to have the employee walk away from it feeling better than when he/she walked in. Start with a list of positive contributions that the employee has made and how they have positively affected you, the group, and the larger organization. For many employees, believing that they are making a positive contribution to the larger organization is more important than any rankings or pay increases. After you have discussed the positive aspects of performance, it is much easier to discuss the areas that need improvement, which is a much nicer way to describe them than to call them negatives or poor performance factors. It is important for employees to know that you take these areas seriously and that you expect improvement. One of the best ways to communicate that is to discuss ways during the review session to achieve the improvement. While you do not have to arrive at the precise solution, it does help to discuss possible training, reading, or other work the employee can do to make the improvements.

4. Give the employees an opportunity to recommend changes that would improve their work experience. Write down what they say and then follow up on it. If they suggest procedural changes that are under your control, follow up and communicate to them what you have done to change the procedures. If they complain about larger organizational practices or policies, pass those concerns along to your management or to your personnel department and communicate to the employee how and when you have done so. It is important for employees to feel as if their manager at least tries to make things better, even if they know that the manager does not have to power to do so. And, even if you can't make the changes happen, it is still important for the employees to have a chance to discuss their frustrations and their ideas for making things better.

5. Close the session on a positive note, again referring to the contributions that the employee makes to the organization and your gratitude for them. Even if you've had to rate someone as "average," it still helps him/her to

feel more positive about his/her job if they know you recognize his/her positive contributions.

The Performance Review Meeting Outline Form provides a template for a performance review script.

Reward and Reinforcement

The nature of technical communication projects often means that TCs do not produce final deliverables for months or, in the most extreme cases, for years. However, employees need and deserve to get feedback, both positive and negative, more often. A technical communication manager, therefore, must seek opportunities for providing such feedback on a more regular basis. The manager must also consider two things related to reward and reinforcement: (1) what will be the nature of the reward and (2) what will be the timing of the reward.

The Nature of Rewards

Do employees prefer money in all circumstances or do other, more symbolic rewards sometimes work better? As the old adage goes, do you give them the turkey or give them the money? A few years ago, the STC asked questions on one of its salary surveys that elicited responses indicating that a majority of technical communicators would enjoy recognition and appreciation for their work more than they would salary increases. This result shows that managers need to find more ways to offer positive reinforcement to technical communicators. For communicators, we hope that a manager can find a way to give both the turkey and the money. That is, we hope that we can reward the employee with recognition and with money. Because of the unusual nature of technical communication work, managers should find ways outside of the formal, structured organizational reward systems to recognize employee accomplishments. While it is nice if the overall organization has annual bonus and team award programs, communicators need to have their work recognized more often.

A technical communication manager, therefore, should include in each annual budget sufficient funds to cover a separate reward program. How big that program should be depends on the nature of the larger organization's programs.

The following sections offer suggestions for reward methods, moving from the smallest, non-monetary feedback methods to those with the largest financial rewards.

Verbal Reinforcement

You can give employees verbal reinforcement at any time and in brief doses. As Blanchard and Johnson suggest in their classic, *The One-Minute Manager* (1992), you can even do so in time periods as brief as one minute. You should ask to be included on the deliverable draft review lists for all of the people who report to you, so that you receive a review copy of each draft they produce, even if you don't have time to read all of them and offer comments. You can look over the drafts to see that they meet your group's quality standards (especially important with newer employees) and to see if the employee has done anything especially innovative

Technical Communication Management Worksheet

Performance Review Meeting Outline

Year _____   Name _____   Date _____

Positive Contributions:

Include the employee's positive contributions to projects, to you, to the group, to your organization, and to the larger organization.

1. _____
2. _____
3. _____
4. _____
5. _____

Areas for Improvement:

Include areas where the employee failed to complete annual objectives, failed to perform adequately on projects, needs to learn more about a technology or a set of tools, etc.

1. _____
2. _____
3. _____
4. _____
5. _____

Employee's Concerns, Ideas for Improvement, Feedback to You, Etc.

Give the employee an opportunity to offer suggestions about how to make the working environment as good as possible. You can also ask for performance feedback about your own work.

Reiteration of Key Positive Contributions

Conclude by repeating the two or three most significant contributions that the employee made this year.

or creative. This gives you a chance to provide feedback, both positive and negative, about their work. Where and when should you do so?

Giving Negative Feedback

Negative feedback should be given as quickly as possible, while the employee still has time to make the changes that your comments will require. Obviously, it should be done in private. Start the conversation by pointing out something positive about their work, something that you liked. Then work your way to the areas where you believe there are problems. This approach helps to prevent the employee from feeling ambushed, and it helps maintain a coaching approach to supervision, which most employees appreciate, rather than a dictatorial approach. You should give an example or two of the types of changes and corrections that you think need to be made because many people have difficulty understanding verbal instructions about a text.

It is a good idea to stop before giving negative feedback and to examine your motivation for doing so. As Latting (1992) points out, managers often give negative feedback "...to control, to express aggression, or to justify your actions." (p. 425). Until you are sure that your motivation is to improve the performance of the employee, you should refrain from giving the feedback.

Giving Positive Feedback

You can give people positive feedback in private too, but it is often more effective to do so in public, even if it is in a small forum, such as a project team meeting or a group meeting. There are many political land mines to consider when giving positive feedback. If you are giving that feedback to one person who is on a team with four other people who are doing the same work, the other four may feel varying degrees of jealousy and that you are "playing favorites." In other words, while it is nice to give people public recognition, you might want to consider giving it privately in some cases. When you do give it publicly, you should state exactly what it was for and, if possible, show examples. That helps to show other employees what your expectations are for high quality work.

Group Recognition

Another effective reward system is to allow anyone in the group to nominate anyone else for an award. In fact, such an award system is most effective when the group members design the program and the awards criteria. Such a system avoids the main problems with many awards programs that are management mandated—that they seem arbitrary, that managers don't know who is really doing the work, that the self-promoting people get the awards, and that managers simply give awards to their favorite employees, their "pets." All of that goes away when the team members choose who should receive awards. The awards can be financial (in which case the manager must provide a budget for the program) or symbolic.

Employee of the Month

You can ensure that at least twelve times a year someone in your group receives an award by giving one out at a monthly staff meeting. This award can be both symbolic and financial. I used "The Cool Idea of the Month" for a while, where an employee who had done something especially innovative received a group coffee mug and a $200 gift certificate for a dinner for two (or whatever else they wanted). Such an award allows both public recognition and some financial reinforcement. One problem with it is that the same employees tend to do the most innovative things, so a manager has to be careful to spread the awards around. On the other hand, any award that everyone gets ceases to become a reward and very quickly becomes an expectation. The award also gives the opportunity for a manager to communicate what he/she thinks is valuable, high quality work. If possible, the manager should show the work that was deemed innovative and explain why, so that group members learn about expectations.

Team Awards

Work today is becoming increasingly project oriented. As Peters puts it, "You are your projects" (Peters 1999). In such environments, it makes sense that awards should be given to project teams as well as (or instead of) individuals. If a technical communication team has achieved some goal that goes beyond normal expectations, its members should be rewarded. There are several ways to do this. One is simply to recognize them publicly and to give them each a symbolic gift along with a financial one. The symbolic gift works best when it is unique and refers to the project specifically—something with the product logo or some other customization. One standard problem when giving team awards is whether to give everyone on the team the same amount or give varying amounts depending on each person's contribution. The latter is preferable, as employees know who has done most of the work and believe that management is clueless if they give everyone equal amounts, even the contractor who came in and helped for the last week.

Another gift that you can give a team is time. At the end of a project, the team may have worked significant overtime. Take everyone to lunch at a very nice restaurant on Friday, and send them all home when lunch is over. And people who have worked extensive overtime can be given a day or two off, unless your organization has a policy against comp time. In that case, violate the policy and give them the time off anyway, calling it something other than comp time (field research, maybe).

Management Bonuses

Many organizations have bonus policies that allow managers to make significant awards for employees whose work has exceeded normal expectations. In scientific and technological organizations, these bonuses often go only to scientists, engineers, and programmers who have done outstanding work, received a patent, won a national award, etc. However, technical communicators should also get their share of such reward systems. Technical communication managers should nominate their outstanding employees for these bonuses and should contend for bonus pool funding. In some organizational cultures, this will require several years of trench warfare, but the end result, having technical communicators included in the overall reward system, is worth it.

Annual Raises and Bonuses

Likewise, communication managers should argue for the same kinds of annual raise and bonus pools that their peers in other departments get. One good way to do this is to bring in the most recent copy of the STC Salary Survey, which shows communicators' salaries increasing at a rate that exceeds inflation. An organization that gives lower than average raises can expect higher turnover and increased hiring and training expenses. In the information age, the most successful organizations will be those which employ experts in designing, developing, and disseminating information. A technical communication manager must educate, cajole, and persuade management to rank and compensate communicators accordingly. Part of that job is education, which can be accomplished using salary surveys, future projections for hiring, etc.

Many organizations determine how much of an annual raise or bonus each person will get based on performance results and on the organization's success in meeting its financial and other goals. If, however, the organization does not, the manager is again faced with deciding whether to compensate everyone equally with bonuses or raises, or to reward them individually based on performance. Here, managers should know that how they apportion raises communicates as much or more than how they conduct performance reviews, both written and oral. The behavior and activities of the people who get the biggest raises and bonuses will be perceived as what you consider important. And don't think that employees will not compare notes about their raises—they will, often down to the last dollar. So you are sending a message in the way you give out raises and bonuses. For annual raises and bonuses, apportionment should be based to some extent on performance, but it is not reasonable to give all of the funds to the top two or three performers while leaving others, especially those whose performance was quite good, with little or nothing. Giving people raises that are less than the inflation rate also sends them a message—that you do not highly value their work and that you would just as soon see them leave. The odds are good that they will. It is better, then, to give annual raises and bonuses in a more equitable fashion, with everyone getting enough to keep up with the cost of living and with better performers getting more.

Promotions

Promoting employees is one of the most difficult yet most rewarding aspects of management. While it is difficult to choose which employees deserve promotions, it can be very rewarding to watch former reports succeed at jobs with increasing responsibilities, both inside and outside of the technical communication field. Hiring, mentoring, and promoting a good employee is very satisfying personally, and it also makes a good, positive contribution to the larger organization. Promotion here refers to advancement to a new job title or level within an existing job category, say from Information Developer 1 to Information Developer 2, or from Technical Communicator to TC Project Manager. While raises are often received when one moves up a salary band within a given level, this section deals with promotion to another level requiring increased responsibilities. Salary increases are usually given as a reward for previous performance, while promotions are given for that reason and for many others, as the following discussion will show.

There are two broad categories of promotions. The first includes merit or proficiency promotions, which are based on attaining a certain skill/knowledge level. This usually means that someone continues to do the same job they have been doing but at a higher pay grade or salary level. Usually their job title changes also, say from Technical Communicator I to Technical Communicator II, or from Junior TC to Senior TC. This type of promotion system assumes that anyone who has acquired the skill set necessary for a certain level should be promoted to that level, and that there are no limits on how many people can be promoted. It also assumes that the person to be promoted has completely mastered the skill set required for his/her current level and that he/she has further demonstrated at least some of the skills for the position into which he/she will be promoted.

The second promotion category, need promotion, fills the organization's need to staff a certain position. Here, the candidate is promoted into a higher-level position that has been advertised and that has a single job available, the one being advertised. Some organizations require all promotions to be done this way, so that rather than being based on proficiency they are based on the number of open positions available at a given level. In this system, a manager must post an open job at the higher position and then allow all interested candidates to apply for that job. Often the manager has a particular person in mind to promote, which leads to the perception that job postings are frequently "wired." In either system, managers must take several things into consideration before promoting an employee.

Promotions are a rare resource for a manager. Because of this, it is important that they be made for the right reasons. Managers use many criteria for deciding whom to promote, including:

- Performance
- Seniority
- Suitability
- Retention/Morale
- Politics
- Favoritism

It is worth examining each of these in some detail, before proceeding with a discussion about how to promote someone.

Performance. This is the most logical criterion for promotion and the one managers most often use (or believe they are using). However, using performance to make promotion decisions is not as simple as it first appears (Hackos 1989). Broadly speaking, there are two ways to measure performance, (1) against a defined set of tasks, skills, and responsibilities (usually formulated in a job description), and (2) performance against implied technical communication measures for excellence, whether they are written down or not. The problem lies with the second choice because as it allows for multiple interpretations by multiple people. One manager may value timeliness over everything else, even document quality. Another might emphasize performance that results in well-written prose. Another might focus on political skills. Peers and direct reports may value criteria other than those that the manager uses for promotion decisions. They will assume that the manager has used some less noble criterion, usually favoritism.

It is far better to promote people based on how well they have performed against a well-documented set of job skills, preferably recorded in a formal job description so that everyone knows what the criteria are. And even then there is plenty of room for interpretation of how well someone has performed the job. The manager's perception may vary considerably from employees who were passed over. Further, is performance at one's current job always the best criterion to use in deciding who will do the best job at the next level? The famous Peter Principle (Peter and Hull 1969) comes into effect here; it states that in a large organization everyone will eventually be promoted into a position for which he/she is not competent. Most technical communicators have had little or no training in how to manage a group (Anderson 1994), but instead have spent their education and post-education training efforts on becoming better communicators. It is entirely possible that someone who is a less capable communicator might make a far better manager than a peer who has superb communication skills. A manager must consider that possibility when deciding whom to promote.

Seniority.   Many organizations have established set levels of promotion after certain periods of service. Sometimes these are mandated. After three years the employee moves from level one to level two, after another five years to level three, and so on. In other cases, period of service is one of several criteria used for promotion. In that case, the person must have worked at level one for three years before being promoted, but performance, suitability, etc. must also be considered, so that promotion is not automatic. In other cases, there is no formal policy, but there is a general sense that one must have served some time in a position before being ready for the next level.

In a skills-based profession such as technical communication, seniority is a lousy criterion to use for promotion decisions. Younger employees often chafe at seeing more senior people promoted whose work they know is inferior to theirs. As the profession moves increasingly into creating information that goes online, on screens, and other media, younger employees with superior skills in those areas will not abide seeing less skilled employees promoted over them simply because they have hung around longer doing "old" technology. Unless it is unavoidable, technical communication should eliminate seniority as a criterion for promotion. In some cases seniority might be a valuable criterion for promotion, especially in large organizations where knowing people in other departments and having a good sense of how the entire organization operates are important assets. . In such cases, technical communication managers should take seniority into consideration. However, formalized, seniority-based promotion systems cause many problems in a technical communication operation.

Suitability.   One of the problems with seniority-based promotions is that people get moved into a position whether or not they have demonstrated the ability to do it. For most technical communication promotions, from one non-management level to another, employees have already demonstrated some of the skills for the next level before they are promoted into it. They have, for example, been given the opportunity to perform project management for an important assignment and have shown that they have good skills. If the next level calls for higher involvement with project management, we are safe in promoting them to it. Things get more difficult, however, when people are being promoted into supervisory or managerial positions. Usually, they have not had previous assignments that required them to perform personnel or budgetary management and they may not have the skills or knowledge to do so. We have all seen examples of the superb technical communicator who consistently produces outstanding documents but who we know would make a horrendous manager. On the other hand, we have seen examples of people whose documents are not as good, but whose experience and knowledge of project management, estimating, budgeting, scheduling, and dealing with people are excellent. Sometimes, then, sheer performance excellence is not a sufficient criterion for making promotion decisions. We must also take into account how well the candidates are likely to perform at the level to which they are being promoted.

Retention/Morale.   Sometimes managers promote an employee because the employee is likely to leave if they don't. In the most extreme cases, the employee comes to the manager and says that he/she has found another position and that he/she will resign unless the manager meets the salary offered by the other organization. To meet that salary, the manager may have to promote the person so that their new salary will fall into the proper band. At other times, managers make pre-emptive strikes, promoting a star before someone else offers that person a better position at a better salary. This works if the person deserves the promotion based on the normal criteria for getting it. However, it does not work if the person is promoted merely for the sake of retention when their skills do not match those of others at the same level.

Politics.   Yes, sometimes promotions are made for political reasons. The manager's boss especially likes a certain employee and wants that person promoted regardless of other criteria. Another reason might involve organizational politics. The person gets along especially well with people in other departments and the new job requires considerably more interaction with them. So we might promote that person even though others fulfill most of our promotion criteria better. Or we might have been allowed to hire the person away from another internal group with the understanding that we would promote him/her, even though others might deserve it more. There are many reasons that political promotions occur. In general, of course, it is best to avoid them and to use more equitable criteria, especially those related to performance and suitability.

Favoritism.   It is human nature to get along better with and feel more comfortable with some people rather than others. Likewise, it is human nature to want to reward those people with whom we interact more easily, especially if promoting them means that we will be interacting with them even more. Some managers are naïve enough to believe that those whom they like more must also be better performers. Those managers who function without much self-examination might even labor under the misconception that it is more important for people to get along well with them than it is for them to demonstrate good performance skills or suitability for functioning effectively at the next level.

Managers who are considering promoting someone should stop for a moment and question whether they are doing it based on established criteria or based on favoritism. You can be sure that some employees will assume that promotion decisions are based largely on favoritism, especially those who are passed over. You also want to avoid the same phenomenon that occurs during hiring, where managers tend to hire people like themselves. Make sure that you aren't promoting the people who most resemble you.

Promoting a Technical Communicator

Promotion practices differ widely. In some organizations, managers have complete authority to promote people through all levels up to positions that are supervisory or managerial. In others, even a move from entry-level positions to the first step up must be approved by management and/or the human resources department. Frequently, internal promotion or management committees must approve all promotions. Getting someone promoted can require extensive effort by the manager. In general, the process involves five main steps, which will vary depending on the nature of your organization. Whether you want to promote someone for merit or need reasons, you will need to perform the following steps.

Choose A Candidate for Promotion

If you look at the reasons we promote people, you can see that deciding whom to promote can be difficult. For merit promotions this is usually not a problem; either the person has reached the requisite skill level or not. But for promotions to fill an organizational need, things become more complex. You will need to prepare a list of criteria for the new position and for filling the need that the organization has. You then have to match the possible candidates you have to those organizational needs. As discussed above, the best performer may not be the best person to fulfill your needs because there may be someone who is more adequately prepared to do so.

If you have someone who deserves a merit promotion, you must decide when to award it. Some managers rarely make a promotion until an employee asks for it. This is a mistake. A technical communication manager should classify all of his/her employees each year (during performance evaluation preparation is a logical time) and decide who, if anyone, he/she wants to promote based on proficiency.

Prepare and Mentor the Promotion Candidate

For those employees who are nearly ready for promotion, you should help them become fully qualified. This may include improving their job skills, adding some project management skills to their repertoire, teaching them about estimating, etc. You can send them for appropriate training but you can also coach them yourself, which is especially effective if they are performing the new skill on the job. You are responsible for making sure you do not contribute to the Peter Principle syndrome by promoting someone who is not ready. Improving the candidates' skills has a two-fold advantage: (1) it helps them get ready for their potential new position, and (2) it provides you with support for your proposal to promote them.

While it may be nice to give someone a surprise promotion, it is usually better to discuss it with them in advance. This is particularly true if the new position moves the candidate from a technical communication job into a supervisory or management job. Some people are simply not interested in supervising. Other factors can also influence someone's desire (or lack thereof) for promotion. I once had someone tell me that she was going through an especially contentious divorce and that she did not want the new position title or income until the divorce was final. I had other people tell me they were simply not interested in supervising people.

Conduct A Public Relations Campaign for the Candidate

In many organizations, promotion decisions are made by management teams or by promotion committees. A manager who wants to promote someone must submit a rationale to the committee to get the promotion approved. Prior to doing so, however, the manager may have to engage in some public relations work on behalf of the employee. As Molisani (1999) points out, public relations activities can include announcements of awards in the organizational or departmental newsletter or Web site, nominations of the candidate for bonuses or internal awards, giving one of your own group's awards to the candidate and publicizing it, having the candidate make presentations to upper management along with you or in your stead, etc. A promotions committee which has seen a candidate's name frequently in a positive light is much more likely to give approval than if no one on the committee has ever heard of the candidate.

The same type of campaign is necessary if you want to get yourself promoted (St. John 1990; Molisani 1999). Many employees resent it as they watch others get promoted but often they have no one to blame but themselves. If your boss doesn't know that you want to be promoted, and if you haven't done anything to help make the promotion happen, it is not surprising that you get passed over. While many employees naïvely assume that management should somehow sense that they want to be promoted and that their work merits it, the smart ones communicate to management their desire for the promotion and make sure that those who make promotion decisions know about their work and their accomplishments. To seethe in frustration while simultaneously refusing to do any kind of public relations to change the conditions causing the frustration is an exercise in futility.

Recommend the Candidate for Promotion

Methods for actually nominating someone for promotion vary widely. In the simplest case, you simply send a notice to human resources announcing that you are promoting the candidate from position x to position y. More often, though, you will have to propose the promotion to your management peers, bosses, and/or to a promotion committee. This proposal should be written or presented, as all good proposals are. It should start with an executive summary, state the candidate's name and the position for which you are proposing him/her, support the promotion with evidence based on qualifications and achievements, discuss alternative candidates and why this one is the best choice, offer a list of benefits derived from making the promotion, and explicitly ask for the committee's approval.

You should also do any of the other work required to get the committee to approve, including attending one of their meetings to defend your choice, politicking with committee members, filling out all necessary human resources organization forms, and whatever else is necessary.

Assist the Newly Promoted Employee

If the people you have promoted stay in your group, you can provide them with better assistance and mentoring in their new position. Whether they stay or not, it is your responsibility to make their transition into their new position as smooth as possible. You may need to introduce them to people in other departments, to project teams, and to new managers with whom they will be working. You may need to help them get additional training as quickly as possible in areas of the job where they are not adequately prepared. And you should be there to provide them with answers and advice when they encounter new problems that haven't been encountered in their earlier positions. If the person does not do well in the new position, it will reflect poorly on you and your organization, so you want to help him/her in any way possible, even if he/she no longer reports directly to you.

Promotions or Transfers into Jobs Outside of Technical Communication

Technical communication is a very diversified profession that touches on many others. Technical communicators often learn skills related to their jobs that allow them to perform other jobs in other disciplines, such as programming, human factors, usability testing, system design, marketing, advertising, management, and more (Maggiore 1991). It is only natural that some employees who learn these other skill sets decide that they would like to practice them full time.

What do you do if members of your group come to you and want to apply for promotions into another department? Or even worse, if they want to apply for jobs outside your organization and they want a recommendation from you? Usually people who have the qualifications and the desire to move on are among your better performers, people who you may badly need to finish their current projects and to contribute to the well being of your group in the future. Some organizations will allow you to cite "needs of the business" and to block internal transfers and promotions, at least for a while. Don't do it. Instead, help the person in every way you can to get the new position. The rest of the people who work for you (and those who might consider working for you) watch these situations carefully. If you help your employees to procure better jobs, whether inside or outside the organization, employees assume that you have their best interests at heart and that you would be a good manager to work for. If you block people in your group from leaving or advancing, they assume that you are motivated more by your own self-interest than by a desire to help those who report to you.

Another reason to help people leave is that it behooves you to have as many people as possible throughout your organization who came from your group and who have positive feelings about you, your group, and the type of work you do. It can also be a benefit to have such people at outside firms where you might be able to arrange partnerships, vendor relations, and the like.

In any case, it simply makes good business sense to help your employees obtain whatever career goals they seek, even if those goals do not involve you and it makes especially good sense in terms of your relations with the rest of your direct reports.

Recommendations and References

Employees frequently ask for recommendations for other jobs, for promotions, and for applications to college programs. You will also get calls from employers who are interested in hiring someone who used to work for you and who has listed you as a reference. When you have positive things to say about the person in question, this is usually not a problem. But what happens when you cannot say much that is positive and, in fact, you are more inclined to say negative things about the employee?

If the employee is applying for an internal position, this is especially difficult. On the one hand, you want to get rid of a problem employee. But on the other, you do not want to give another group your problem, and you especially do not want to do so by exaggerating the employee's skills and professionalism. For internal references, it is far better to be totally honest, especially if there are performance problems associated with the employee. If the employee is looking for a position in another department, you can be candid about his/her work within our discipline and offer your opinions about how well the person may perform in some other discipline. But you do not want to overstate the employee's qualifications just to get rid of your problem and to give it to someone else who will certainly hold it against you if the person does not work out.

Some organizations, for legal reasons, do not allow managers to make recommendations or give references. If you say negative things about an employee and prevent the employee from getting a job, you can be sued, even if the things you said were demonstrably true. Conversely, if you say highly positive things about the person and the new employer finds them not to have been true, they also can sue you (and this happens!). So, many organizations will only give prospective employers the dates of employment and job titles of former employees.

A manager must decide what his/her policy will be here. I was willing to violate organizational guidelines and to give positive references about employees whose work I had respected. When I could not give a positive reference, I simply stated that organizational rules prevented it and referred the callers to the human resources department, who would offer only dates and titles.

Giving positive recommendations and references can be tricky. So many of them are filled with superlatives that readers have come not to trust them very much. It may be preferable to write them in a more understated fashion, which can actually engender more trust than the ones that sound like movie advertisements.

Reductions in Force

Reducing staff can happen for many reasons, including a general business downturn, a slowdown in a specific industry, or ill fortune for a single organization. In any case, technical communication managers may occasionally have to participate in reductions in force, sometimes euphemistically labeled "laying people off," even though the chances of recalling them are low to non-existent.

What criteria do you use to determine whom to "reduce?" Sometimes organizations make this easy for managers by designating the criterion to be used: e.g., last-in-first-out, lowest on latest performance review, everyone at a certain level, etc. Often, though, the manager is asked to prepare a list of employees ranked from lowest to highest for priority on the layoff schedule. This is, of course, a heart-wrenching exercise. The most prominent criteria used include seniority, performance, and estimation of future success.

Seniority

Perhaps the easiest method to use in deciding whom to terminate is a reverse seniority system, simply laying off the most recently hired employees. While this does provide the benefit of simplicity and the appearance of fairness, it has two major problems. First, you may lose the vigor, new ideas, and experience with emerging technologies that newer employees often bring, which can be vital for a technical communication organization. Second, while it appears fair, it is in fact not fair to people who believe that criteria other than seniority, especially performance, should be used for layoffs. The victims and the survivors will see that more capable employees are being sent out the door while less capable but more senior people remain.

Another layoff criterion, once used often but now used rarely, is to lay off all of the senior people. This is compelling because you save more money faster. You might also get rid of some "dead wood," people whose productivity has slowed and who are not as enthusiastic about new ideas as they once were. This method also has two major problems. Losing all of the knowledge that more experienced employees have can seriously affect how well an organization operates. Further, it is against federal and state laws to lay off older employees first. Most larger organizations are aware of such employment laws and will carefully screen layoff lists to ensure that they are not violating the laws and leaving themselves susceptible to large discrimination lawsuits.

Seniority, then, has significant problems as a reduction criterion. Unless mandated by one's organization, it is probably not the wisest choice.

Performance

Another criterion often used is performance. Those who earned the lowest ratings on their latest performance reviews or, if formal reviews are not done, who are deemed by their managers to be the lowest performers are the first to be laid off. Some organizations regularly engage in small reductions in force of 5 to 10 percent simply to force managers to get rid of the "dead wood." Their upper managers believe that getting rid of non-productive employees balances the lost productivity and lower morale that results from reductions (a dubious tenet). Paradoxically, the biggest advantage and disadvantage of using performance as a layoff criterion is fairness. Some see using performance as the only fair way to proceed, so that the strong and capable are being rewarded while the weak and incompetent are being "weeded out." On the surface, this system seems more fair than any other.

The difficulty with using performance comes in people's perceptions of the manager's assessment of each employee's performance. This is especially true in organizations that do not have a formal performance management system, where no criteria are published to evaluate performance. The lack of such standards makes any assessment by the manager susceptible to charges of favoritism, discrimination, and incompetence. If my performance is rated low, I am inevitably going to charge that the manager arbitrarily plays favorites and rewards the toadies and the "yes" people while failing to realize my own brilliance, intelligence, creativity, and productivity. No matter how assiduously a manager follows sound performance review methods, some people are going to perceive a level of unfairness and arbitrariness associated with any such system. Some of the remaining employees are going to resent the fact that one or more of their friends were laid off, even if they know in the back of their minds that those people did not perform as well as those who stayed.

Perhaps the biggest problem with using performance as a criterion comes when a group is populated by employees who are all solid performers, where there are simply no weak links. In such a case, the performance differences may be so miniscule that using them to determine layoffs indeed seems arbitrary and unfair to almost everyone.

Despite its problems, performance is a compelling criterion to use for forced reductions. For technical communicators, who engage in an intellectual and creative exercise, performance seems to be the fairest method to use.

Estimation of Future Success

Organizations undergoing rapid and dramatic changes may choose to reduce force by assessing which employees they estimate will be most successful in a changed working environment. This is particularly true in technologically oriented organizations, where change can often be the only constant. Many technical communication managers have had to contend with such change as the traditional paper documents their groups developed were being replaced with on-line delivery, single sourcing, and information management systems. Those communicators who were willing and able to change from the paper mentality to the on-line mentality were coveted employees. However, those who were not willing to make that transition may no longer be productive contributors. When an industry rapidly changes its processes and its products, some employees can make the switch and some simply cannot or will not.

For some technical communication managers, then, assessing the future suitability of employees for new working procedures and outputs can be an effective and desirable method to use if reductions are forced.

Other Reduction Considerations

Another aspect of staff reduction is the consideration of employees' personal situations. Whom does one lay off first: a single, working parent who is not receiving any child support payments or the spouse of someone who earns six figures in a "safe" industry? What if the most popular person in your group, the one who helps keep people working hard and morale high, is the person who shows up at the top of your layoff list, based on the criteria you are using? What if someone who has "seen the light" and made tremendous improvement in the last year or two in skills and performance still doesn't quite measure up to someone who has been around forever, not improving but working at a reasonably high level? Some would argue that you should stick with a single criterion regardless of other considerations. However, for the ongoing benefit of the group, it may be worth making an exception when special circumstances dictate it.

Overall, there is no single system for deciding whom you must lay off. For technical communicators, some mixture of performance and assessment of future success seems to be the most reasonable method, with consideration given to any special cases.

Problem Employees

Because technical communication as a field does not have a licensing or credentialing system, anyone can claim to be a technical communicator. Strangely, many people believe that they are "good writers" when in fact they are not. People who would not dream of claiming to be auto mechanics or brain surgeons feel perfectly free to claim that they are technical communicators, even though they may not have any more credentials or experience writing about science and technology than they do repairing automobiles or performing brain surgery. For that reason, there are people in the profession who are basically frauds and with whom an unfortunate manager may have to deal. Obviously, the way to avoid this is with a careful interviewing and hiring process, but some of the frauds are going to slip through even the best screening systems.

Everyone who supervises other people will eventually have to deal with employees whose performance is not satisfactory. This is one of the most difficult jobs for a technical communication manager because in our discipline the quality standards are so varied and difficult to measure. Often, our assessment that someone's work is not meeting requirements will seem to be subjective and susceptible to challenges. Dealing with performance problems is also difficult because of the psychological stress caused by looking someone right in the eye and telling them that they are not measuring up. Nonetheless, managers make more money because they have to do these things. Doing them in the proper manner can remove some, albeit not all, of the trauma for everyone involved.

The most important thing to remember here is that you must focus on the effects of a person's performance problems rather than on the causes. It is natural to want to help someone you see struggling with personal problems, family problems, substance abuse, and other difficulties, but the job of a manager is to assess performance, not to act as an amateur psychiatrist or counselor. This may sound cold, but there are good reasons for it. First, when it comes to someone's personal problems, you are indeed an amateur, and your best intentions may make the problem worse rather than better. Well-intentioned managers have ended up being sued because they tried to offer support in an area where they had no training or expertise, with tragic results. Second, when you set yourself up as counselor to your employees, you establish a dangerous precedent. If you help one person with an extreme problem, do you have to help others with milder problems? Employees can very quickly figure out that their performance shortcomings will be overlooked if they present their personal problems to you. Another reason to focus on performance problems rather than their causes is the legal thicket involved. You cannot force an employee, as a condition of employment, to get counseling for substance abuse or for personal problems. Forcing the person to put up with your counseling could have serious legal ramifications. In serious cases, you should seek the assistance of human resources and, if your organization has one, of your employee assistance professionals.

A manager must consistently focus on the performance of employees rather than on the causes for that performance. You cannot make it a condition of employment that someone stop drinking. You can make it a condition of employment that his/her performance must achieve a certain level of quality and quantity. Rigorously focusing on performance also helps the employee to see directly the consequences of their behavior, rather than wallowing around in the causes.

Most performance problems are not caused by such serious matters but are largely a matter of training and motivation. The following process will help you deal with an employee having performance problems, up to and including termination in the unfortunate case that it becomes necessary.

Communication

The first and foremost rule when dealing with performance problems is to communicate, communicate, communicate. The most important communication is with the employee directly. If matters get worse, communication will become increasingly important with your boss, human resources personnel, and, perhaps, your organizational legal department (Gerencher 1999).

Most performance problems are fixed by simply discussing them with an employee and pointing out how they negatively affect the goals of your group and of the organization overall. Employees usually want to perform well, so most of them will try very hard to correct problems and to improve their performance. While it may sound Pavlovian, it is best to communicate about a problem as soon as possible after it happens. The impact is more likely to affect the person's performance than if you wait six months until their next formal review. Communicating performance problems should consist of three elements:

1. A statement about what the problem is
2. An analysis of how improvements can be made and what the proper outcomes should be, with examples (if appropriate)
3. A statement that you expect those outputs to occur by a certain date

With technical communicators, the most common performance problems that warrant a performance meeting are deadline problems, drafts that do not meet your quality standards, and teaming problems with other communicators and/or with subject matter experts.

The deadline problems are the most straightforward here. You owe it to the employee to listen to their reasons for why they have missed a deadline, which will usually be based on someone not supplying information to them. However, it is important to clarify for some communicators that you expect them to actively seek information, to attend project meetings, to read specs and requirements, to use prototypes, etc. to get the information they need rather than to passively wait for a subject matter expert to hand it to them. A performance problem meeting, however brief, is not a time for too much empathy. You are trying to communicate to the employee that you are dissatisfied with his/her performance. With some people, this requires very direct, focused communication that leaves no room for ambiguity. If you make qualifying statements, some employees will hear only those and will assume that they are off the hook.

Communicating quality standards is indeed a difficult business in technical communication. You are lucky if your organization has published standards to which you can refer. With no published standards, it becomes more difficult to hold employees responsible for quality, other than some amorphous concept that you hold in your head. At the very least, your organization should have some standards for layouts, organizational identity, abbreviations, etc. If not, assign someone to work on an organizational or project style guide. Even with published standards, though, there is still plenty of room for employees to make errors in their documents, whether in overall organization, technical accuracy, usability, or other areas. One of your main responsibilities as a technical communication manager is to establish standards in those areas (written or not) and to communicate those standards to employees. Brief performance problem meetings are one of the ways to do that. In such a meeting, you sit with the employee, go over the document and point out the problems, suggest ways to fix the problems so that they meet standards, and request that the employee do so by a certain date. You should not have to go over every single problem with the document, but rather one example of each type of problem. A good technical communicator should be able to take that information and to revise the document satisfactorily.

As technical communication becomes increasingly on-line and multimedia oriented, more and more collaboration with teams of variously skilled employees becomes mandatory. The days of the lone technical writer working on a document are numbered (although they will probably never completely disappear). Unfortunately, some communicators much prefer this older model. They may have been very successful using it (including awards, good reviews, and nice raises) and may be reluctant to move toward a more collaborative working model. Or they may simply be so introverted and shy that they just don't want to work with other people. Your job as manager is to establish collaboration guidelines and expectations, and to communicate them clearly. It is important to note that you should not discuss someone's personality here (even if it stinks). Rather, you should discuss the performance results of how he/she interacts with others, and you should tie those results to their impact on the project and the organization. You should show the employee how his/her behavior is affecting outcomes that are important and explain to him/her how to modify the behavior so as to contribute to the desired outcomes rather than detracting from them.

Documentation

With most performance problems, one or two meetings with an employee, even brief ones, will clear up the difficulty. However, if the problem persists, or if the employee is unwilling or unable to correct it, you must begin to document what is going on. As things get worse, more frequent and more formal documentation is required. You hope you do not have to move up the documentation hierarchy because of performance problems, but it is a good idea to use that hierarchy when it becomes necessary (Garron 2001). While the following hierarchy will vary among organizations, it represents common practice:

Informal Feedback Sessions.   These "one-minute manager" (Blanchard and Johnson, 1982) meetings allow you point out a problem to the employee and suggest ways to fix it.

Performance Meetings.   Use either scheduled or ad hoc meetings with the employee to go over a performance problem, review expectations, offer examples of acceptable fixes, and request that changes be made by a certain date. At this point you might not be documenting in writing, but you should start doing so if you have to have more than one or two of these meetings with an employee.

Memorandum to File.   Write a memorandum and put it in the employee's personnel file when you have observed a continuing performance problem. These are appropriate when you don't have time to meet with the employee or when you already have explicitly laid out your expectations, and observe that they continue not to be met. It is a good idea to copy someone else on this memo, so that you can prove that you wrote it on the date when you say you did.

Memorandum to Employee.   As a follow-up to a performance meeting, write a memorandum to the employee to explain the problem, your expectations for it to be solved, and the date by which that should be done. This more formal communication indicates that you are moving toward taking serious action related to the employee's performance problems. You should probably get human resources and your boss involved at this point, and they should certainly be copied on this memo.

Letter to Employee.   This is a formal letter detailing how performance expectations have not been met and perhaps stating that continued employment may be in question if those expectations are not met. This is usually sent only if you believe that the employee is not going to improve and that you are going to have to fire him/her.

Action Plan.   An action plan is a last-ditch effort to "save" an employee. It also gives a manager the justification for terminating an employee. The action plan states explicitly what objectives the employee must meet and defines the quality expected. It also includes specific dates by which each objective must be met. It further states that if any of the objectives are not met at the required quality level by the required date, that the employee will be terminated as of that date. Usually, plans are developed with human resource department assistance and with the approval of one's manager. See the section below on Action Plans for more detail.

Termination Letter.   This is a letter given to an employee on the date of termination. How explicit the letter is about the causes for termination is usually a call to me made by human resources and legal personnel. Assuming that sufficient documentation exists in the employee's file, the letter can simply state that the employee is being terminated for cause effective on a certain date. It should also state whether the employee will be given a two-week notice period and pay for that period.

Rarely is it necessary to go through this whole process, especially if one has communicated adequately with the employee early on. In some cases, though, it may be unavoidable.

Action Plans

Keeping an employee is usually preferable to termination, which involves trauma, involves much time and paperwork, invites potential legal challenges, and requires the time and expense of hiring a replacement. Action plans give managers several advantages when they are trying to keep an employee. First, the plan spells out quite explicitly for the employee the objectives and the dates by which they must be met. An action plan helps educate the employee as to the realistic productivity and quality expectations for a professional technical communicator. It also communicates the urgency of learning about such things because it usually includes a date at which the employee will be terminated if the objectives are not met. At the very least, it provides a manager with documentation of unsatisfactory performance if terminating the employee becomes necessary.

Preparing an action plan requires several steps:

1. First, make sure you follow all organizational guidelines and the letter of the law in preparing the plan and in dealing with the employee.

2. Meet with your human resources department and your supervisor so that they are aware of and can participate in preparing the plan.

3. Prepare a realistic set of objectives and tasks for the employee to complete in some fixed time frame, usually three months or six months.

4. For each objective, spell out in as much detail as possible what you expect the outcome to be, that is, precisely what the nature of its deliverable will be. For example, if one of the outcomes is a first draft by a certain date, specify what that draft must contain (90% of text completed, graphics included in at least sketch form, table of contents but no index, etc.). Include a table that shows each objective, the specific outcome required to fulfill that objective, and the date by which it must be done. These should be routine objectives that you would normally give to a technical communicator as part of their job. The idea here is not to trick or trap someone. Rather, it is to impress upon the employee the normal expectations for a technical communicator and to determine, under very controlled circumstances, whether the employee can meet those expectations.

5. Include a section at the end of the plan that explicitly states that failure to fulfill all of its objectives will result in termination by a certain date.

6. Meet with the employee, go over the plan, taking as much time as necessary to answer questions and provide even more detail. The employee may complain that the expectations are excessive, but you must maintain that they are consistent with what is expected of other employees and are what is required for continued employment in your organization.

7. Write a memo detailing the date and time of the meeting and summarizing your discussion in the meeting. Attach a copy of the action plan and put it in the employee's personnel file.

8. Conduct weekly meetings with the employee to track performance against the action plan.

For an example of an action plan, see the Action Plan Worksheet.
When you present an employee with an action plan, you are likely to experience one of four possible outcomes:

1. Some employees will simply resign. This saves you the considerable headache and paperwork of going through the termination process.

2. Some employees will admit that they have not been doing adequate work at the job they are in and will ask for assistance in finding some other type of work. See the following sections for how to help an employee in this situation.

3. Some employees will do everything in the plan exactly as spelled out, even though they think the whole thing is a game and they do not change their attitudes about work or about making a real contribution. When the plan dates are over and you have retained them because they succeeded, they go right back to the kind of performance that forced you to prepare the plan in the first place. When this happens, you write another action plan.

4. And, sometimes, employees will realize that they have not been performing adequately, will work hard to correct their problems, will successfully complete the plan, and will continue to perform at a satisfactory level afterward.

Obviously, the last possibility is the one that we would prefer, although there are times when the first one may come as a relief. Because technical communication does not have any entry requirements, we occasionally encounter people who have called themselves technical communicators and who, perhaps on the basis of one or two relatively easy early writing assignments, have concluded that they indeed are experienced technical communicators. When they are faced with more difficult, complex assignments, they may lack the training, knowledge, and/or aptitude to complete them successfully. When this happens, it is in a manager's best interest to help employees find other positions (internally or externally) where their knowledge and abilities fit better. The following sections describe ways to do so.

Aptitude/Attitude Tests

Technical communication managers occasionally encounter an employee who does not have the qualifications to do quality technical communication work but who promises otherwise to be an effective employee. Also, a technical communication manager occasionally meets with a communicator who wants to do something else. Some employees may have decided, after a few years, that technical communication simply isn't for them anymore, or they may have decided, perhaps from project work they have done, that some other aspect of developing scientific and technological products is more to their liking. They may also have decided that they want to develop communication products for some other industry. In such a case, you can refer them to Lutz and Storm's *The Practice of Technical and Scientific Communication: Writing in Professional Contexts* (1998), which covers a broad range of technical communication jobs, providing examples and information related to the nature of the work, the requirements for entry into the field, sample salaries, and more.

The best thing a manager can do here is to help the employee in every way possible. If your organization's human resources department provides testing services for aptitude, suggest that the employee take the Myers-Briggs or other similar survey to help them in making career choices. While the Myers-Briggs and similar tests yield data about a person's preferences and attitudes, a newer testing method yields information about a person's strengths. See Buckingham and Clifton's *Now, Discover Your Strengths* (2001), or go to http://www.strengthsfinder.com.

If your employee cannot take such a test internally, offer to pay for testing by an external career counseling organization. The few dollars this will cost will be considerably less than you will spend trying to prove that the person is not a competent technical communicator. Many aptitude and attitude tests are now available online at moderate cost. Taking some of these tests can help a burned out or marginally competent employee decide about other career options. You are much better off if employees leave your group happy and having felt supported than if

Technical Communication Management Worksheet

Action Plan

Name _____   Date _____

You are being required to work under this Action Plan because your work performance has not met expectations as described below (and in your performance review of day/month/year).

[Describe the results of the employee's performance; i.e., missed deadlines, poor quality, failure to work well with team members, etc.]

You must complete each of the objectives below at the level of quality specified and by the date specified. Failure to complete any of the objectives can be cause for termination.

Objective	Quality	Level Due Date
Deliver draft one of the XYZ User Guide	90% text 100% graphics in sketch form or better Fewer than three technical errors as determined by SME and managerial review 100% conformance with organizational style requirements, per the ABC Style Guide, as determined by the project editor	June 30, YYYY
Objective 2		
Objective 3		

If you have not adequately completed each of the above objectives by June 30, YYYY, your employment with ABC will be terminated, effective on that date.

Signed:

_____   _____

Manager   *Department Director (and/or HR rep)* you simply throw them out the door. Plus, other employees will be watching how you handle the situation, and seeing that you are trying to help the employee having problems will improve your reputation with them. Granted, you do not have to do anything in this situation but measure the employee against your performance and quality standards and discard them if they do not measure up. But you will come out ahead in the long run if you provide assistance. It is far more humane and less costly to have people walk out the door happily than to force them out. Further, it is more in line with what employees would expect from principle-based management.

You can also help employees by looking for other opportunities for them internally where their skills and interests might be better employed. While a manager should not foist his/her problems off on someone else, if you have an employee who shows strong aptitude for and interest in another discipline, it makes sense to help that person find opportunities there.

Assistance with Job Search

Likewise, it also is good business practice to help such employees look for jobs outside your organization. This includes networking with people you know who might have openings in appropriate jobs, helping the employee look for job postings online, including those on STC and other sites related to our field. Some employees made the right decision to join our discipline, but they may have opted for the wrong niche within it. If the person has strong graphic design skills but does not write well, for example, you can help find a position either internally or externally that will better use his/her skills and interests.

While it is inappropriate and unethical for a manager to write glowing recommendations for an employee whose performance has been mediocre, there is nothing wrong with making factual references and recommendations to other managers about someone who has shown true aptitude and preference for a particular type of work. And again, it is far less expensive and more rewarding for everyone involved to help employees find more appropriate jobs than to simply fire them.

Termination

If you have tried to help an employee in every way possible and you have had the person complete an action plan with less than satisfying results, you may have no choice but to terminate their employment with your organization. In 15 years as a manager, I terminated only one full-time employee. Many others resigned or moved to other departments after they were put on action plans or after I helped them find more appropriate positions. Termination, then, is truly a last resort when someone simply cannot do the job or will not modify his/her behavior so that his/her performance and quality reach satisfactory levels.

I am discussing here termination for performance and quality reasons, not terminations that result from single or multiple offenses that warrant immediate removal, such as criminal activity, theft of organizational property, sexual harassment, violence, or gross insubordination.

A technical communication manager should not undertake alone firing someone for incompetence. Rather, he/she should have the support throughout the process of the internal human resources department and of his/her supervisor. This is necessary to insure that other eyes than the manager's look at the evidence gathered and agree that it warrants dismissal and that the manager has handled the case in as fair and impartial a manner as possible. Conflicts between an employee and his/her supervisor often occur more due to personality differences than to inadequate quality and quantity of work. It is easy for a manager to make mistakes in such a circumstance and to try to make a case for dismissal when there really isn't one. Hence, it is important for others to participate in the process. Another reason for involving others is to get their counseling about what steps to take next and how to carry them out so that the organization has minimal legal vulnerability. This can also require getting the organization's legal department or lawyer involved.

Several steps normally occur when dismissing someone:

1. Document everything that happens near the end of the process. That includes saving all e-mail correspondence with the employee, notes on conversations, and memos to file regarding the person's performance.

2. Working with human resources, your supervisor, and, if necessary, legal assistance, write a letter to the employee informing him/her that as of a certain date his/her employment with your organization will cease. The letter should not make general remarks about competence, but it can refer to previous communications such as performance reviews, action plans, memos, etc. The letter should state explicitly when the person should be out of the office and when pay and benefits will cease. Because of the ability of a disgruntled employee to do major damage to the software files that we use to prepare documents and on-line systems, I prefer to require a terminated employee to leave immediately, even though we will continue to pay them for two weeks.

3. Set up a meeting with the employee. Get your supervisor or a human resources representative to attend the meeting also, so that you have a witness to what is said in the meeting.

4. Decide whether the termination is immediate or the person will have two weeks (or some other period) prior to leaving. I recommend that technical communication terminations be immediate, as a fired employee can do much damage to file systems and to the morale of others by being a strongly negative force while still on site. The humane thing to do is pay him/her two weeks severance pay but send him/her home immediately.

5. If you believe that the person is capable of destroying computer files in a pique of anger before leaving, arrange with your network administrator to backup appropriate files on the day of your termination meeting and ask the administrator to block access to the computer at the scheduled time of

the meeting. This may sound harsh, but a knowledgeable person can do considerable damage to local and network files in only a few minutes if he/she is angry enough.

6. Notify your organization's security personnel that you will be terminating someone at a certain time and place. This is a precaution that is probably not necessary but, given the varying reactions of people to being terminated, it is a good idea to have your security people in the loop.

7. Meet with the employee, read through the termination letter, present it to him/her, and inform him/her that he/she has 30 minutes to pack up his/her personal possessions and leave. Inform him/her that he/she must not take any organizational property, either physical or intellectual, including files or documents on which he/she has been working.

8. After the employee leaves, notify your group, either in a meeting or via email, that the employee has left. Tell him/her how you will handle getting the work done that the employee was handling. Resist the temptation to go into any detail about why the person was fired. If the ex-employee learns that you have been making negative comments about him/her publicly, you and your organization can be sued. You simply tell people that the employee is no longer with the organization, and nothing more.

9. Prepare yourself for the psychological effects the termination will have on you. Firing even someone who richly deserves it is still not an easy thing to do. Stay busy, go home, cook your favorite meal, and listen to your favorite music.

Teamwork and Collaboration

Team Building and Collaboration

As most technical communication work becomes increasingly team-oriented and collaborative, a technical communication manager must know some of the methods to enhance good teamwork skills and to avoid serious conflicts among team members. Given the nature of technical communication work, that collaboration often occurs over a distance, and a technical communication manager must know how to manage people in multiple locations and how to build a team even when some (or all) of its members are not co-located.

Nearly all technical communication is, in one way or another, collaborative. Even a lone technical communicator at an organization must work with SMEs and other reviewers, which means that effective collaboration must occur. As technical communication moves more toward delivering a diversity of products (beyond paper books), the very nature of the job becomes more team-oriented and collaborative. For example, developing an on-line, multimedia Web site requires site designers, system engineers, writers, graphic artists, programmers, usability testers, and others. No one person can have sufficient expertise to perform all of the required tasks at a sufficient level. Hence, technical managers increasingly must assemble and work with a diverse collection of job skills and professionals who perform them (Ede and Lunsford 1990; Lay and Karis 1991; Spilka 1993; McGarry 1994).

Managers must work on two aspects of team building and collaboration. First, the manager must handle the people aspect, selecting team members, communicating project planning and goalsetting, facilitating communication and interaction, and encouraging positive conflict while preventing negative conflict (Bosley,1991). Second, the manager must handle the physical aspect of team building, including the facilities where people work and meet, the tools they use for collaborating, and hardware and software infrastructure necessary to support effective collaboration (Burke,1992).

Personnel Aspects of Team Building and Collaboration

By far the most important managerial challenge for building teams with diverse employees who have diverse duties is to employ sound leadership techniques, especially related to giving everyone the big picture. If employees are completing their tasks in isolation and are unaware of the overall project, chaos ensues when disparate components, reflecting varying conceptions of how things should work, are finally brought together. If, however, employees all see the larger picture, they share a common goal and a common conception of what the final product should look like.

This is a good reason to have the whole team involved in the initial project planning, rather than doing it yourself or delegating it only to a project manager or senior technical communicator. The more that team members see the whole picture, the more likely they are to work together effectively and to create a communication solution that meets the goal.

Building an effective team requires selecting the appropriate team members, both in terms of skill sets and temperaments. If you put two or three control freaks together on a team, you are asking for trouble. Each of them will try to take control of the project, usually with disastrous results. If you put only one control freak on a project, you can count on that person doing everything possible to take it over. On the other hand, assigning such a person to handle the project management aspects of the work that others may not want to deal with is not a bad idea. If you look for people who are quiet and non-assertive, you may get a team that appears to be functioning very well but that is not doing much that is innovative or pushing toward more effective solutions. Technical communication requires real, creative effort. Creating is a messy business that often cannot be forced into a nice, pleasant, predictable mold (Hackos 1990; Hansen 1988). An effective team that is creating something should have some conflict. The ideal members of such a team are people with strong opinions who are willing to state them assertively and then to negotiate toward a common conception that includes as many of the best ideas as possible.

On a technical communication team designing and developing new (or heavily revised) texts, you need several types of workers:

Innovators, creators, visionaries—people who can see the big picture, who can look at the audience and its tasks and then design the optimum library of communication solutions to meet their needs. These people are often not good at doing the other kinds of work listed below.

Implementers—People who can take the concepts of the innovators and turn them into real-world plans, with specifics on numbers of screens or pages, types of graphics, etc., and who can then create texts that match those plans.

Project Managers—People who keep the project running, who can estimate the hours needed to prepare the proposed library and then track progress toward completing it, in terms of both time and dollars expended.

Workers—People who do what needs to get done, who help in any way necessary to get the texts completed.

Producers—People who are good at moving texts from draft stages to final output stages, whether this entails formatting paper documents or writing code for on-line documents.

Finishers—People who get the product out the door, who take care of the last minute details involving final production, indexing, translation, printing, and, sometimes, shipping.

Obviously, there is much overlap among these groups, but a technical communication project needs someone who has the knowledge and the inclination to perform each of these roles. A technical communication manager should choose team members with that in mind and, ideally, should assign the various roles to team members at the outset of the project.

Any new team is going to go through a period of what the team-building experts call storming. During this period, team members vie for power and for having their opinions and ideas heard. Some of this is natural and desirable. However, a manager must ensure that the team gets beyond the storming stage with everyone still working toward the final goal and without anyone getting miffed and exercising pocket vetoes, that is, undermining the project plans even though he/she "agreed" to them.

To some extent, the success of getting a team to work together effectively depends on the personality types assembled. While many people assume that personality trait surveys, such as the Myers-Briggs, are bunk, many others have found that having all of the members of a team take such tests together and compare notes is an excellent team-building exercise (Leonard 1993). The process of doing the task together may be more important than the actual test results. Certainly a by-product of having a team do such an exercise is that team members learn that other people prefer certain communication methods, have differing value and reward systems, and have different strengths. One excellent choice is Buckingham and Clifton's *Now, Discover Your Strengths* (2001), and the accompanying Web site, http://www.strengthsfinder.com. This test provides information about each employee's five main strengths. Taking the test on-line and then comparing information about strengths should provide a team with a better idea of how to apportion work and with an increased respect for one anothers' abilities and interests.

Such awareness helps people work together more effectively whether the test results themselves are meaningful or not. Another method to help build teams is to have the team members attend a team-building class wherein they perform various tasks together. Such exercises, offered by many training firms, help a team get through the storming stage, as team members develop roles and learn about others during the class rather than doing so in the more pressured environment of the workplace.

Another good practice for improving team collaboration is for team members to get to know one another outside of the office. This can be as simple as having the first team meeting at a restaurant at lunchtime. It can also involve a half-day or full-day meeting at an off-site location, which might be someone's home (possibly the manager's), a hotel meeting room, or a rented restaurant room. It is also a good idea to have the team engage in some social activity together, one that is not work-related and does not entail an official team meeting. One group I was in went to the racetrack together and all made $2 bets on the ponies. Another went bowling together, which proved to be a universally humiliating, hilarious, and bonding experience. It is, of course, a good idea to choose activities that everyone on the team can engage in.

Another choice that teams must make is in the model of collaboration they will follow. On the one hand there is the old, waterfall model of project management that assumes that various tasks are completed by various functional groups and then handed off to other functional groups who perform their various tasks before making another hand off, and so on. As Wambeam and Kramer (1996) point out, such an approach often leads to isolation, little collaboration, and failed projects. They propose what they call the dialogic model, in which team members work together throughout the project. This approach has the advantages of having more good minds work on each aspect of the project, of having an overall context for the project that everyone is working in, and of improving intra-team communications throughout the project.

Physical Aspects of Team Building and Collaboration

A technical communication manager needs to do everything possible to remove physical restrictions to successful collaboration. The ideal here would be for the team to be located together in a perfect working environment. Because this rarely happens, the manager needs to help use physical and technological tools to approximate that ideal as closely as possible. While it is not my purpose here to discuss the specific tools available now, it is worthwhile to discuss the concepts behind the different types of tools. This list is not comprehensive, but it should provide a manager with a starting point for developing the appropriate physical tools for good team collaboration.

Having a project team located together within a cubicle area or in adjoining offices obviously fosters better team communication. Office environments with some flexibility in rearranging workstations and equipment can dynamically maximize the configuration, enabling project teams to grow and shrink and to remain co-located.

Other technologies are important for helping teams work together, including conference telephone systems; on-line project repositories and communication systems; and electronic communication systems including computer networks, e-mail, instant messaging, list servers, and special collaboration software such as Lotus Notes and various Web-based products, often called "groupware." Teams working collaboratively need communication that is as reliable and immediate as possible, and a technical communication manager should institute as many technological solutions as are necessary to ensure such communication. Networks are almost a necessity for effective file sharing, graphics transfers (the large file sizes often cannot be sent via email), and general collaboration.

Communication tools are especially important when some or all of the team members work in different physical locations. In such cases, teleconferencing, on-line conferencing, e-mail, and instant messaging become even more vital.

Conflict Resolution

Any team is going to have conflicts. Some conflict is healthy, as it means that team members are offering different ideas and debating their merits. This kind of substantive conflict (Burnette 1991) is natural with technical communicators, who may arrive at different interpretations of audience analyses and the appropriate solutions for meeting the audience's needs. So long as such conflicts stay positively focused on arriving at the best solutions, the conflicts should be welcomed. However, when they move away from ideas and toward personal enmities or arguments about processes, they force a manager to respond quickly. Allowing conflict to fester can lead to numerous problems, "choosing sides," lower productivity, lower quality, and even sabotage of others' work efforts. As soon as you become aware that two employees have a conflict, you should act.

Communication is key. The first step is to meet with the two people and allow each of them to tell his/her side of the story, while the other listens without interrupting. Such a meeting will allow you to determine how serious the conflict is. After both people speak, you can ask them for their ideas about solving the problem. Ideally, you will arrive at a negotiated solution, put it in writing, implement it, and follow up by a specified date to ensure that the solution has worked (Andes 1999). However, you may also have to counsel one or both of the employees in private sessions about the realities of workplace trade-offs and compromises regarding documentation solutions and project processes. Further, you may have to send one or both of the parties for training related to whatever is causing the problem. While it is important to try not to make one person feel like the loser in the debate, it is equally important to resolve the conflict and get the parties focused on working productively again. This may mean that you have to choose between proposed ideas for documentation solutions or for work processes. It is far better, though, to work with the two people so that they believe that they have arrived at the conclusion rather than having you impose it on them. As Leonard (1993) points out, solving conflicts in such a collaborative manner can have powerful positive benefits for a technical communication team.

Collaboration with Scientists, Engineers, and Developers

Successful technical communicators not only collaborate well with other communicators but also with the technical people with whom they work. In the most uninformed and unsuccessful model for such collaboration, communicators work in isolation from the development team, often not joining a project until it is near or at completion. Obviously, technical communicators should be members of the development team from its inception, and they should work closely with the developers throughout the design and development effort (Bresko 1991; Wambeam and Kramer 1996; Colvin and Beecher 1999).

Technical communicators often report considerable difficulty communicating effectively with engineers and scientists, commonly known as subject matter experts (SMEs). The most often cited problems are the SME's lack of time for or interest in collaborating, hostility on the SME's part toward working with a writer, misunderstanding by the SME of the task-oriented nature of user documentation versus comprehensive technical description of the product, frustration of the SME by the communicator's lack of technical knowledge, or belief by the SME that documentation is unimportant because "no one reads the manual anyway." Walkowski (1991) finds that technical experts also view the collaboration as difficult, and cites their concerns about communicators' technical knowledge, writing and language skills, communication ability (including oral communication), attitudes, and professionalism.

As DeGraw (1993) points out, most articles about getting along with SMEs say "… you should win the developer's respect, establish a firm schedule, avoid being hostile, and do your homework before interviewing someone." (p. 80). While these steps are a good start, they hardly represent real collaboration and teamwork. Such collaboration requires a more concerted effort, including several key concepts.

Technical communicators should avoid the "us versus them" mentality and terminology. Bryan (1994) describes the "Us-Them" social dynamic that seems to occur in almost any social group. Unfortunately, for many technical communication groups, the "them" are the very developers and/or scientists with whom technical communicators must work successfully to accomplish their goals. It is easy for this polarity to develop, as communicators and engineers/scientists often come from differing educational backgrounds, have different discourse communities, hold different beliefs about which parts of a product or service are most important, and frequently compete for coveted project and organizational resources. Hence, the two groups whose collaboration is key to producing high quality products often work together reluctantly and, in the worst cases, with open hostility. One of the first ways to avoid this polarity is to avoid identifying the groups with significantly different labels. Rather than calling the technical experts "SMEs," call them "co-workers" or "co-developers." Consider labeling the technical communication group with a term that more closely fits within the organizational milieu than does "technical communicator" or "technical writer." IBM, for example, has long called its communicators "information developers." On one of my jobs, I was a "supervising proposals engineer," which fit well with the engineering staff with whom I had to collaborate. A technical communication manager should avoid criticizing developer/scientist groups and should encourage employees to avoid doing so. Rather, the manager should encourage employees to view technical co-workers as people with whom we share a common goal and with whom we must collaborate successfully to achieve that goal.

Another important way to foster collaboration is to get to know developers/scientists as people and not merely as information sources (DeGraw 1991; Sopensky and Modrey 1995). This requires conversing with them in non-work-related situations, such as social gatherings, water cooler bull sessions, lunches, intramural sporting events, organizational picnics, and other venues. While collaboration can succeed among people who barely know one another, it obviously has a much better chance to succeed if those people are more closely affiliated and share some common interests. A technical communication group can often foster these relationships by sponsoring activities at work that are fun and that allow people to interact, even briefly, in a non-work-related environment. And communicators should actively participate in organizational events where they can interact, rather than stay sequestered in their own segregated enclave.

Finally, and most important, technical communications professionals should manage the relationship with technical experts. If left to chance, the relationship, given the cultural differences between the groups, is not likely to develop well. Managing the relationship requires several steps:

1. Ensure that effective experts are assigned the responsibility of working with communicators. Managers in scientific/engineering groups often have no idea who is most effective in working with communicators, so their assignments of such responsibilities are often not based on whether there is much chance for success. A technical communication manager should communicate often with peers in other organizations about which experts work well with communicators and which ones do not. The technical communication manager should provide performance feedback to experts' managers so that those who work effectively with communicators are rewarded. The technical communication manager should also specifically request that those who have proven most effective be assigned to collaborate with communicators while those who are least effective and even hostile should not be assigned such duties.

2. Train technical experts and reviewers. Most scientists/engineers have no idea how to mark up a document effectively to ensure technical accuracy and maximum usability. Often, technical personnel have been taught that documentation means comprehensive explanation of every facet of a product's operation. Indeed, programmers are often taught to "document" their code, which means to include a comment about almost every line explaining the function of that line. Hence, their idea of documentation is that it should explain in painstaking detail every function of a product. A training session can help explain to them that documentation intended for customers must focus on customer tasks and not product tasks and features. The training session can further explain effective methods for marking up review drafts. For example, the trainer can show that simply writing "no" in the margin of a paragraph does not tell the communicator what is wrong and requires the communicator to take up more of the expert's time in finding out what the problem is and how to fix it. Thus, it is more efficient if the

expert provides at least a brief explanation of the problem on the mark-up. Such a training session can be called a coordination meeting and need not take more than an hour or two, but it can help ensure more valuable reviews and better working relationships.

3. Communicate dates and milestones well in advance. This is simply sound project-management practice (Hackos 1994), but it is often not followed. Technical experts should have plenty of advance notice as to when they are going to receive review drafts and when they are expected to return them.

4. Prepare a set of responsibilities so that the experts and the communicators can define their roles in the collaboration (Shouba,1999). This can include expert responsibilities such as returning drafts on time, being available for a certain number of hours a week for consultation, providing requirements documents and technical specifications, etc. For communicators, it can specify obtaining as much information as possible without requiring expert input, establishing project delivery/review dates and milestones in advance, keeping drafts up to date with project developments, and other duties.

5. Prepare a review cover sheet for every review draft sent out. The cover sheet should include the name(s) of the reviewer(s), the communicator(s), the date when the review is due, places to write comments, and places where reviewers can sign off that they have conducted the review.

6. Communicate progress regularly at status meetings and with written or e-mailed communications on a regular basis (weekly or monthly, or both, as the project schedule requires). If a project schedule is threatened because of slow review returns, the reviewers should be notified first. If they continue to miss deadlines, their management must be notified. In any case, a technical communication manager must constantly communicate progress to avoid being left holding the bag at the end of a project when everything else is on time but the documents are late.

7. Give feedback to experts and their management. Effective reviewers should be rewarded, at least with a memo to their management (and copied to them, if appropriate). An employee who sees management in another organization sending positive feedback to his/her manager is likely to work hard in the future to maintain the positive relationship with that other organization. If you are making team awards to an especially effective technical communication team, include the most effective experts in those awards. Unfortunately, it is also necessary to communicate to experts' management when one or more of them have not proven to be effective in working with communicators. This should be done in a non-accusatory tone and in a manner that does not put the expert's manager on the defensive. It can simply be pointed out that the person did not work as effectively with communicators as others do and that perhaps the person needs some training or the technical manager might consider assigning documentation responsibilities to someone else.

Technical communication managers should not passively assume that they must accept whoever is assigned SME responsibilities. It is the communication group's goal to work successfully with scientific/engineering personnel to accomplish organizational goals. Collaborating effectively with technical personnel is critical to achieving those goals. Such collaboration is a two-way street, however, and communicators should not simply accept ineffective SMEs. Open and frequent communication with technical management and effective management of the collaborative effort can considerably improve the working lives of communicators and the success of their projects.

Other Technical Communication Personnel Issues

Technical communication managers must deal with many personnel issues other than those directly related to hiring, mentoring, promoting, and terminating employees. One important consideration involves ergonomics, especially considering the often extensive hours communicators spend working on computers. A related issue involves workplace flexibility and telecommuting, given that many communicators can do at least some of their work at home. Another important personnel-related issue concerns how to handle personal leaves, child care accommodation, and the like.

Ergonomics

Technical communicators spend something like 30 to 50 percent of their time writing. The rest is spent in meetings, interviews with customers and SMEs, testing product prototypes, etc. Nonetheless, there are periods where technical communicators spend many hours sitting at their desks typing on a computer. The technical communication manager should do everything possible to make workstations as ergonomically effective as possible, for two reasons. First, doing so reduces the health problems that frequently afflict technical communicators, including eye strain, back problems, and repetitive-motion problems. Second, better ergonomics lead to higher productivity, as comfortable and healthy workers produce more and better texts.

While it is not within the scope of this book to present a full ergonomics course, technical communication managers should learn about the ergonomics of office configurations, especially as they affect those who work on computers for long hours. At the very minimum, technical communicators should have chairs with adjustable heights and armrests. They should have computer monitors with at least 17-inch, non-glare screens, with 19-inch and higher being preferable, particularly if they need to read the equivalent of two 8 $\frac{1}{2}$ x 11 inch sheets on the screen at once. They should have keyboards with adjustable heights and with wrist supports to reduce carpal tunnel syndrome. Lighting should be adequate for reading paper texts comfortably, but not so bright as to cause screen glare. Managers should aggressively pursue funding to support the best ergonomics possible for communicators. While it may be part of the macho milieu in some scientific and technological firms that engineers and programmers work amidst squalor in unsafe, non-ergonomic environments and get the work done anyway, communicators who have to function in such environments are eventually going to have health related problems, which wind up costing much more in lost productivity, health-care benefits, disability pay, and, in some cases, lawsuits.

Work-style Flexibility for Technical Communicators

A wise technical communication manager pays more attention to the quality and quantity of someone's work as opposed to when it is actually performed. Technical communication is creative work. Some creative people work better at 3:00 A.M. than they do during regular business hours. While it is important for communicators to attend project meetings and to work with other communicators, SMEs, audience members, and others, there is a significant part of the work that must be done at one's computer. Often that work is done more efficiently without other people around and without the normal interruptions of the daily workplace. I once did a time inventory study (see Chapter 6) with an employee in which we counted the phone calls, e-mail chimes, and office visits received during the average day. This person was interrupted, on the average, every seven minutes. For many communicators, it is difficult if not impossible to write effectively with interruptions that frequent. For this reason, technical communication managers should allow their employees wide latitude in when and where they get the designing, creating, composing part of their jobs done. That includes telecommuting for one or two days a week, working off-site for extended periods, and working non-standard hours for extended periods. It also involves trusting employees to get the work done by certain deadlines, without being concerned about when they do the actual work or if they even appear at the workplace for several days. I had one employee who would disappear for many days in a row, sending me e-mails about his progress at various bizarre hours, and then reappear with excellent work.

The wise technical communication manager also learns that, while we can expect people to work extensive overtime, sometimes even for extended periods, we cannot and should not expect it on a permanent basis. While the work-around-the-clock mentality of many technological organizations makes for an interesting mythological, romantic milieu, it is also bad business. It leads to less productive employees, higher burnout and turnover rates, more mistakes, higher absence rates due to physical and mental problems, and products of lower quality. Salaried employees should not get or expect compensatory time, but it is a good idea to give someone who has just put in six 75-hour weeks to get a project finished two or three days off, with pay. It is also a good idea to plan off-site meetings, picnics, trips to the park, and the like well in advance and have the whole group go, no matter what is happening with their schedules. This not only builds teams, but it gives employees some needed rest from the ardors of everyday work and, perhaps, the opportunity to step back and look at what they have been doing from a larger perspective.

In any case, technical communication managers need to learn to be very flexible regarding how and where their employees work. Some employees will be more productive and happier if they are allowed to do much of their work off-site and at odd hours. Some will want the assurance of the 8:00-5:00 workday with colleagues around them. In either case, the technical communication manager needs to help the employees be as effective as possible.

Telecommuting

The nature of technical communication work makes it not only possible but in fact desirable for at least some writers to telecommute. There are several advantages. First, we are contributing to a cleaner environment by requiring fewer people to drive to a central office. Second, we are accommodating good employees who might otherwise have to leave. Third, telecommuters who are working on some aspects of technical communication (especially writing drafts) can be considerably more productive when working at home. And fourth, we are able to form teams that simply could not be put together without some members telecommuting full-time (Weber 1996).

A technical communication manager should work with the human resources department and upper management to determine telecommuting policies, including:

- Do you provide equipment or is the telecommuter expected to supply it?
- Do you pay for their network access from home?
- Do you provide and pay for a second telephone line?
- Are telecommuters covered by your organization's insurance while working at home?
- Who is responsible for technical problems with the telecommuters' computer and communication hardware and software?
- Are telecommuters expected to be instantly available while working at home? This is significant. In some systems, telecommuters are held responsible for responding immediately to telephone and electronic communication. In other words, they are in effect working the same 8:00-5:00 hours that other employees are, but they are doing it at home. In other systems, they are held accountable for getting a certain amount of work done, but they do not have to respond instantly to communications. The manager should determine which system to apply and should communicate it very clearly to telecommuters.
- How will their work be saved, backed up, and secured? How will it be protected regarding proprietary issues?

Once these issues have been resolved, a manager and a telecommuter need to develop an understanding, perhaps in writing, as to how the telecommuter will function. Will he/she be responsible for being immediately available or for getting the work done during a 24-hour (or weekend) period? Will he/she be responsible for attending meetings via teleconferencing or other means on the days when telecommuting? Which meetings require the person's physical presence and which can be attended remotely? Putting all of this in a memo that the telecommuter and manager sign helps everyone know what the expectations are.

As Langhoff (2001) points out, managing by results rather than by observation will help a technical communication manager deal with telecommuting. With experienced and conscientious employees, managing telecommuting is generally no more difficult than managing locally. The one difference is that managers may have to break tasks down into smaller components to ensure that telecommuters are staying on schedule, but even this is not usually necessary with experienced employees. It does become important when a manager suspects that a telecommuter is not managing time well and is not getting enough done during telecommuting time. The best way to manage such a situation is to tell the employee explicitly about the concerns and to work with him/her to develop a more detailed schedule with smaller deliverables until he/she proves that he/she can indeed keep to the schedule. While telecommuting benefits everyone involved, managers should not allow it to be used as an excuse for missing deadlines and falling behind schedule. A manager may have to make the hard decision to revoke the privilege of a telecommuter who is not keeping up.

Technical communication managers need to learn how to manage telecommuters and remote workers, as it will be more and more difficult in the coming years to assemble teams of communication developers with all of the needed skills in one geographical location. This means that managers need to develop and sharpen their project management skills. Effectively managing remote workers mainly requires that one have well-defined project goals and deliverables with controlled schedules.

Leaves, Day Care, Family Issues

The wise technical communication manager will develop a flexible system for allowing employees to take leaves for pregnancy, child care, and other types of family issues. This is true for two primary reasons. First, it makes good business sense to have employees who are working at high productivity and who do not harbor grudges against a job, an organization, and a boss whom they see as taking them away from important family matters. Second, with the shortage of good technical communicators, such flexibility helps retain good employees when finding and hiring new ones is difficult.

Usually, leave and day care policies are determined on the organizational level rather than by individual managers. However, technical communication managers are often well served by showing even more flexibility than the organizational policies would normally allow, even though that may carry some risks. Because of the nature of technical communication work, it may be possible to find types of work that an employee can perform at home over an extended period while he/she is also caring for an infant, a sick child, or an infirm parent. If the manager follows the careful project management practices described above in the telecommuting section, these arrangements can often work. However, even if a good employee wants to take several weeks or months off without pay and beyond the normal organizational guidelines, it is a good idea to try to accommodate him/her. Doing so shows that you care, which is important, and it also helps you keep a valuable employee.

Privacy, E-mail, and Web Surfing

Technical communicators have to do considerable research, sometimes on internal products and services and sometimes on external subjects. Hence, they often use the Internet as a research vehicle. They also have to use e-mail extensively—to interact with other team members, to collaborate with the subject matter experts with whom they are working, to send and receive drafts and comments, to communicate about project progress and problems, and other important functions.

Problems can arise with whether employees have any rights to privacy and whether they should be allowed to use their organizational computers for personal reasons. This is a sticky legal thicket. The courts have ruled that employers own everything associated with an employee's work, which includes the information on the hard drive of that employee's computer. While there are constant challenges to such rulings, they have been consistently upheld. That means that, legally at least, an employer has the right to completely monitor an employee's on-line activities, to read all of the e-mail, to check all of the files on the computer, etc. Should an employee's manager do so? Also, should a manager forbid technical communicators from conducting personal matters on their organizational computers during normal work hours?

This is one of those matters where common sense is the best solution, and where we would prefer to avoid having to develop "official" rules and regulations regarding computer use. Managers are wise to allow employees to conduct a few minutes of personal computer use during the normal workday. If a technical communicator who has been working 12-hour days to get a project in on time wants to order his/her child a birthday present, the manager should happily allow it. If one technical communicator wants to send e-mail to several others inviting them and their spouses over for dinner on Saturday, only a few bits of corporate resources have been expended, and group harmony may have been improved. Unfortunately, some employees cannot restrain themselves from overusing computers for personal reasons, and these people often force organizations to adopt restrictive measures. If possible, though, managers should respect employee's privacy and allow moderate use of computing facilities for private reasons.

Should employees be allowed to surf the Web during business hours? In the case of technical communicators, yes. Increasingly, we expect communicators to develop and deliver information via the Web. One of the best ways they can learn effective Web layouts and navigational systems is to look at other sites. If technical communicators want to wander around on the sites of competitors for a few hours prior to designing a new system for our organization, we should cheerfully permit them to do so. If they want to spend the same hours searching for the perfect green sweater to get Aunt Tilly for her birthday, then we need to caution them about abusing the privilege of using the Web during work.

In all the years that I managed communicators, I had only one person's computer and mail system searched, after some highly inappropriate material was discovered in one of that person's drafts that was distributed for review. The search, conducted by corporate security, showed that the computer's disc drive was loaded with inappropriate material and that such material was being e-mailed to people outside the corporation. It also showed that the person harbored some chilling ideas about the other employees in our group. The person was summarily dismissed.

Despite that kind of experience, I believe that we should permit reasonable personal use of computers. So long as communicators do not abuse the time and the resources of the organization, more good than harm will come from doing so. A communication manager should be able to tell if someone is not getting his/her work done in a timely manner. If that happens, the employee needs a warning that spending too much time surfing the Web or sending e-mail or instant messages is viewed as a performance problem. Otherwise, if people are getting their work done, we are wise to let them spend some time surfing and e-mailing.

References

Alexander, S. 1998. Bringing outsiders in. *Infoworld*, 31 August, : 79–80.

Andes, R. 1999. Don't let conflict get you off course. *Infoworld*, 9 August: 69.

Armbruster, D. L. 1986. Hiring and managing editorial consultants. *Technical Communication* 33(4): 243–246.

Aughey, K. 1999. Manager and leader—a look at the differences. *STC Newsletter of the Management Special Interest Group* 3(3): 4.

Blanchard, K. and S. Johnson. 1982. *The one-minute manager*. New York: William Morrow & Co.

Bosley, D. S. 1991. Designing effective technical communication teams. *Technical Communication* 38(4): 504–512.

Bresko, L. L. 1991. The need for technical communicators on the software development team. *Technical Communication* 38(2): 214–220.

Bryan, J. G. 1994. Culture and anarchy: What publications managers should know about us and them. In *Publications management: Essays for professional communicators*. Edited by O. J. Allen and L. H. Deming. Amityville, NY: Baywood: 55–67.

Burke, A. C. 1992. "Managing ourselves through quality teams", in *STC 39th Annual Conference Proceedings*, Washington, DC: Society for Technical Communication.

Burnette, R. E. 1991. Substantive conflict in a cooperative context: A way to improve collaborative planning in workplace documents. *Technical Communication* 38(4): 532–539.

Carlson, P. 1999. "Information technology and organizational change." *Conference Proceedings, 17th Annual International Conference on Computer Documentation*. New Orleans: Association for Computing Machinery (ACM) Special Interest Group on Systems Documentation (SIGDOC).

Buckingham, M. and D. O. Clifton. 2001. *Now, discover your strengths*. New York: The Free Press.

Colvin, R. D. and V. Beecher. 1999. "Improving the writer developer relationship." *STC 46th Annual Conference Proceedings*. Society for Technical Communication.

Dalla Santa, T. M. 1990. The whys and hows of effective performance appraisals. *Technical Communication* 37(4): 392–395.

DeGraw, Y. 1993. Us vs. them—why can't we be friends? *Technical Communication* 40(1): 80.

Ede, L. and A. Lunsford. 1990. *Singular texts/plural authors: Perspectives on collaborative writing*. Carbondale, IL: Southern Illinois University Press.

Garron, R. S. 2001. Take action on poor performance. *Infoworld,* February 12: 69.

Gerencher, K. 1999. Tackling terminations. *Infoworld*, March 1: 77–78.

Hackos, J. T. 1989. "Documentation management: Why should we manage?" *STC 36th Annual Conference Proceedings*, Society for Technical Communication.

Hackos, J. T. 1990. Managing creative people. *Technical Communication* 37(4): 375–380.

Hackos, J. T. 1994. *Managing your documentation projects*. New York: John Wiley & Sons, Inc.

Hansen, N. J. 1988. "Managing creative employees." *STC 35th Annual Conference Proceedings*, Washington, DC: Society for Technical Communication.

Hopwood, C. V. 1999. Staff training: Filling the gaps. *Intercom* 46(1): 18–21.

Houser, R. E. 1998. Create your personal training plan. *Intercom* 45(10): 4–9.

Horowitz, R. B. 1994. "Improving managerial-employee communication: a case study." *STC 41st Annual Conference Proceedings*, Society for Technical Communication.

Katz, S. 1999. *The dynamics of writing review: Opportunities for growth and change in the workplace*. Norwood, NJ: Ablex.

Langhoff, J. 2001. Managing by remote control. *Intercom* 48(1): 24–27.

Latting, J. K. 1992. Giving corrective feedback: A decisional analysis. *Social Work* 37(5): 424–430.

Lay, M. M. and W. M. Karis. 1991. *Collaborative writing in industry: Investigations in theory and practice*. New York: Baywood.

Leonard, D. C. 1993. Understanding and managing conflict in a technical communication department. *Technical Communication* 40(1): 74–80.

Maggiore, J. 1991. "Technical communication: A training ground for management." *STC 38th Annual Conference Proceedings*, Society for Technical Communication.

McGarry, C. 1994. An overview of collaborative writing for the publications manager. *Technical Communication* 41(1): 26–33.

McDowell, E. E. 1994. "Scientific and technical communicators' perceptions of the performance appraisal interview." *STC 41st Annual Conference Proceedings*, Society for Technical Communication.

McGuire, G. 1991. "Working in 'Flow': A theory of managing technical writers to peak performance." *STC 38th Annual Conference Proceedings*, Society for Technical Communication.

Molisani, J. 1999. Advance your career using public relations. *Intercom* 46(7): 9–11.

Molisani, J. 1999. Tools or talent? Hiring a technical writer. *Intercom* 46(2): 24–25.

Muench, B. S. 1995. "Documentation team leadership in the 1990s." *STC 42nd Annual Conference Proceedings*, Society for Technical Communication.

Peter, L. J. and R. Hull. 1969. *The Peter Principle: Why things always go wrong*. New York: William Morrow & Company.

Peters, T. 1999. The wow project. *Fast Company* 24: 116–128.

Roff, H. V. 1993. "Performance reviews: A chance for two-way communication." *STC 35th Annual Conference Proceedings*, Philadelphia, PA: Society for Technical Communication.

St. John, M. 1990. Market yourself into management. *Technical Communication* 37(3): 366–369.

Schriver, K. A. 1997. *Dynamics in document design*. New York: John Wiley & Sons, Inc.

Shouba, T. K. 1999. Bringing in the SMEs. *Intercom* 46(4): 13–15.

Smith, C. 1993. "Peer mentoring as a means of career development." *STC 40th Annual Conference Proceedings*, Dallas, TX: Society for Technical Communication.

Sopensky, E. A. and L. Modrey. 1994. "A manager's toolkit for hiring the right writer—or how to avoid throwing a wrench in the works." *STC 41st Annual Conference Proceedings*, Society for Technical Communication.

Sopensky, E. A. and L. Modrey. 1995. Survival skills for communicators within organizations. *Journal of Business and Technical Communication* 9(1): 103.

Spilka, R. 1993. *Writing in the workplace: New research perspectives*. Carbondale, IL: Southern Illinois University Press.

Thuss, L. K. 1982. Improving document flow with external resources. *Technical Communication* 29(2): 4–7.

Walkowski, D. 1991. Working successfully with technical experts—from their perspective. *Technical Communication* 38(1): 65–67.

Wambeam, C. A. and R. Kramer. 1996. Design teams and the web: A collaborative model for the workplace. *Technical Communication* 43(4): 349–357.

Weber, J. H. 1996. "Taming a telecommuting team." *STC 43rd Annual Conference Proceedings*, Washington, DC: Society for Technical Communication.

For Further Information

JoAnn Hackos's *Managing Your Documentation Projects* (1994), while focusing on project management, nonetheless explains numerous methods related to technical communication personnel management. Likewise, her article, "Managing Creative People" (1990) discusses the special personnel management concerns associated with technical communicators.

There are dozens of books, articles, and Web sites that treat the general idea of management vs. leadership. The subject is discussed in relation to technical communication by Kathleen Aughey in "Manager and Leader—a Look at the Differences" (1999) and Patricia Carlson in "Information Technology and Organizational Change" (1999).

Norma J. Hansen in "Managing Creative Employees" (1988) and Gene McGuire in "Working in 'Flow': A Theory of Managing Technical Writers to Peak Performance" (1991) discuss leadership methods for technical communication managers, including ways provide employees with greater motivation.

Karen Schriver's *Dynamics in Document Design* (1997) discusses the various ways in which technical communicators (and others) conceptualize communicators' roles, the historical development of such conceptualizations, and, indirectly, methods that will help managers provide appropriate conceptualizations to their personnel and to other groups within the larger organization.

Jack Molisani in "Tools or Talent? Hiring a Technical Writer" discusses assessing job needs and candidate proficiencies when making hiring decisions. John

Bryan treats the acculturization of new employees and other ideas about how technical communication groups relate to their larger organizations in his chapter, "Culture and Anarchy: What Publications Managers Should Know about Us and Them" (1994). Susan Katz's book, *The Dynamics of Writing Review: Opportunities for Growth and Change in the Workplace* (1999) describes how managers can use writing reviews to help acculturate new employees. Catherine V. Hopwood in "Staff Training: Filling the Gaps" (1999) explains methods for developing effective training systems for technical communicators. Rob Houser's article, "Create Your Personal Training Plan" (1998) describes how technical communicators can develop specific plans for their professional growth. Carole Smith gives ideas for technical communication mentoring programs in "Peer Mentoring as a Means of Career Development" (1993).

Performance reviews specifically for technical communicators are discussed in T. M. Dalla Santa's article, "The Whys and Hows of Effective Performance Appraisals" (1990). Blanchard and Johnson in their classic *The One-Minute Manager* (1982) discuss methods for providing ongoing performance feedback to employees.

For employees who are not performing well as technical communicators, standardized aptitude tests may help. The classic is the Myers-Briggs test, which now has many variations, some available on the Web (simply search on 'Myers-Briggs' for many examples). Perhaps more effective and more current is Buckingham and Clifton's book, *Now, Discover Your Strengths* (2001), and the *StrengthsFinder Profile* online test that can be taken after purchasing the book. For more information, see http://www.strengthsfinder.com.

On the subject of team building and collaboration, there are several good sources devoted strictly to the subject of collaboration on technical communication teams. Deborah Bosley's article, "Designing Effective Technical Communication Teams" (1991), gives an excellent introduction to the subject. Three books provide greater detail about building technical communication teams, working collaboratively on such teams, and working collaboratively with subject matter experts and developers in other areas: Ede and Lunsford's *Singular Texts/Plural Authors: Perspectives on Collaborative Writing* (1990), Lay and Karis's *Collaborative Writing in Industry: Investigations in Theory and Practice* (1991), and Rachel Spilka's *Writing in the Workplace: New Research Perspectives* (1993). Carol McGarry provides a good summary of the subject and the research in "An Overview of Collaborative Writing for the Publications Manager" (1994). David Leonard describes methods for managing conflicts on teams, including the suggestion that group aptitude testing may help, in "Understanding and Managing Conflict in a Technical Communication Department" (1994). Rebecca Burnette also distinguishes between positive and negative types of team conflicts in "Substantive Conflict in a Cooperative Context: A Way to Improve Collaborative Planning in Workplace Documents" (1991).

Likewise, several authors have treated the subject of collaborating with scientists, engineers, and programmers. Laura L. Bresko's article, "The Need for Technical Communicators on the Software Development Team" (1991) provides arguments to support early participation by communicators on software teams.

Wambeam and Kramer, in their article, "Design Teams and the Web: A Collaborative Model for the Workplace" (1996) discuss using dialogic rather than formal collaborative systems for technical communication projects, especially for developing Web materials. Yvonne DeGraw in "Us vs. Them: Why Can't We Be Friends" (1993) and Sopensky and Modrey in "Survival Skills for Communicators Within Organizations" (1995) discuss strategies that include becoming acquainted with subject matter experts as people and as friends, including outside of the workplace, rather than merely as sources of information. In an excellent article, "Bringing in the SMEs" (1999) Terryl K. Shouba recommends helpful methods for managing the relationships with subject matter experts.

Jean H. Weber, in "Taming a Telecommuting Team" (1996) and June Langhoff, in "Managing by Remote Control" (2001) provide suggestions for managing technical communicators who are telecommuting and/or working at remote sites.

Questions for Discussion

1. Who is the best leader you have ever worked with? Why? What specifically did he/she do that motivated you?
2. Should technical communication managers hire people based on their knowledge and skills with the tools they use or based on their abilities at audience and task analysis, document design, and writing ability?
3. Have you ever had a formal, written training plan? Have you ever written one yourself? Are such plans a hindrance to flexibility or a beneficial guide?
4. After performing especially well on a project, would you rather receive a plaque honoring your performance or a check for $100? Why?
5. Are employees better motivated by having several grade levels in their function that they can strive for, or are they happier with one level that has a wide pay range and that makes everyone "equal," at least in title?
6. Technical communicators often work for months or sometimes even years without completing a project. How do you motivate people in that position with financial compensation? With other kinds of compensation?
7. Should technical communicators be paid as much as engineers or programmers? More? Less? Why?
8. If you work (or have worked) in a technical communication position, compared to what other professionals earn, are technical communicator salaries greater, equal to, or lower? Why? What parts of the organization's culture and power structure cause technical communicators' salaries to be where they are? If necessary, how would you try to change the salary structures?
9. What team(s) have you been a member of that was highly successful? Why? What team(s) have you been a member of that was quite unsuccessful? Why?

CASE 2

Compensation and Rewards Plan

The Management Situation

As the new manager of the Technical Communication group at Aardvark Enterprises, you are in the process of hiring several new employees into your group. Because the company has not previously had a technical communication group, there is no compensation plan in place for your group. Obviously, you need to put such a plan together before you can proceed with interviewing people and making them job offers.

Normally, compensation planning is the purview of experts in the personnel department, who research similar companies and geographical areas to arrive at compensation plans that will keep a company competitive in hiring and retaining employees. A full compensation plan may be hundreds of pages long and list detailed specifications for every job classification in an organization, including factors such as salary, medical benefits, savings plan matching, pensions, tuition reimbursement, bonuses, deferred tax savings programs (401k), travel and entertainment allowances, transportation allowances, leave policies, part-time pay systems, and more.

You have not been asked to prepare such a detailed plan, as Aardvark's Personnel people will handle most of the details. However, they do not have what they consider to be reliable industry information about technical communicator pay scales, and they want you to recommend what those scales should be. Further, they do not know how many levels of technical communicator you envision for your group, nor how those levels should overlap for compensation purposes.

You will have to consider many variables while preparing your abbreviated compensation plan for Personnel. How many levels do you want? One, with a very broad pay range? Two or three, with narrower, overlapping, ranges? As your organization grows, will you have specialists such as editors, graphic artists, multimedia developers, project managers, trainers, and others? Will you need separate categories for them or will you be able to include them in the levels you define now? What will you call the level(s)?

In addition to the levels and pay scales you propose, Personnel also wants you to propose some sort of reward system outside of the annual raise structure. How will you reward someone who has developed an especially innovative solution, or who has worked extensive overtime for months, or who has developed an especially positive interaction with a key customer? Several years ago in one of the STC salary surveys, they asked members if they would rather be rewarded with money or with recognition. A larger percentage said that they would prefer recognition. How can you set up a reward system that accommodates that preference, but that still includes a financial component? How do you provide positive feedback and reinforcement for people who may complete projects only once every few months?

The Assignment

Prepare a one- or two- page memo to Personnel with your recommendations for a compensation plan for technical communicators at Aardvark Enterprises.

The plan should specify how many levels of technical communicators you propose, what each level should be called, and the salary range for each level.

In addition, propose some type of recognition/reward system, including how much "bonus" money you propose to use each year, expressed as a percentage of overall salaries. Your management is going to review your proposals. They expect you to request enough funding to hire and retain excellent employees, but also to conserve the company's resources. In other words, you cannot simply propose outlandish pay scales for technical communicators, or you will be called onto the carpet about it. You are looking for a salary plan that allows you to recruit and keep some of the better people in your area, but that also falls within company financial and cultural norms.

Helpful Hints

Try some of the following sources for more information about compensation planning.

For a brief introduction to compensation planning, including links to many other sites, see http://management.miningco.com/library/weekly/aa041498.htm?pid=2737cob=home

To get the STC's most recent salary survey, go to the following url: http://www.stc.org/salary.html

Search the nearest STC chapter's Web site to see if they do local salary surveys. To find a list of chapters, go to http://www.stc.org/chapter/chapter_search.asp

Evaluation Criteria for Case 2

Have you developed pay scale(s) that are competitive in the your geographical area? Would you be able to hire people and keep them at the levels you've set?

Have you accommodated hiring of entry-level people? Experienced veterans? Does your proposal allow for future hiring of people in associated disciplines who may not be technical communicators per se?

Have you devised a system for rewards that includes a financial component but also some way of recognizing people's contributions in other ways?

Will your management be happy with your pay scales and convinced that you are marshalling organization funds carefully while still insuring that you can hire and keep good employees?

CHAPTER 3

Finance for Technical Communication Managers

Introduction

If technical communicators are to understand the overall picture of the products and services on which they work, they must understand the overall financial strategies employed by their organizations. A technical communication manager who must make difficult decisions about assigning priorities to project staffing, for example, needs to understand how those projects fit into the overall organization's financial picture.

Many managers in scientific and technological organizations perceive that technical communicators do not understand the organization's overall missions and goals, especially those that are related to finance. This perception, unfortunately, is the reality in many places. Technical communicators prefer to focus on the rhetorical and linguistic challenges they face rather than the financial challenges that the overall organization faces. Most communicators prefer to deal with words and not numbers, and so managers at upper levels and in other departments assume that technical communication groups provide a needed "service" but that they do not really understand what the organization is doing. Technical communication managers must counter this perception so that they can demonstrate how they add value to the organization's mission, can assure some degree of viability for their groups, and can participate more fully in the activities of the overall organization, including how it gives out raises and rewards.

New technical communication managers are perhaps least prepared for budget and finance issues. In their jobs prior to management, they generally do not learn much about financial issues, save working on project estimates, usually for only the part of a project represented by their individual documents. Further, little or no instruction on finance and budgeting is included in most technical communication degree programs because the emphasis is more often put on rhetorical and design strategies. And technical communicators have not often taken any accounting or business courses in school.

This chapter covers the four finance areas that technical communication managers will most often have to confront, including:

- **Balance Sheets and Income Statements.** Balance sheets give an overall perspective of an organization's finances. While usually the purview of upper management and stockholders, they may be used by technical communication managers who are entrepreneurs or who are working in small companies. Income statements show the revenues versus the expenses for a given period, thus yielding a bottom line showing whether the organization has had a profit or loss for that period.
- **Budgets.** Technical communication groups are funded in many different ways. Some TC managers are not responsible for budgeting, while some are held totally accountable for their group's budget, both on a project-by-project basis and an overall basis.
- **Estimation.** All technical communication managers will do estimates. If not, they will rue the consequences of having someone else do them and then impose the estimates, no matter how unrealistic, on the technical communication group.
- **Value Added.** Technical communication managers should be able to demonstrate how they add value to an organization's products, processes, and overall mission and goals. This section includes a formula for calculating value added.

Balance Sheets and Income Statements

Balance Sheets

Balance sheets show the overall financial picture of an organization, including its assets, its liabilities, and its capital. While it is beyond the scope of this book to present an entire Business 101 course, some preliminary definitions are necessary before a balance sheet will make sense to a technical communication manager.

Assets—the group of resources owned by an organization. There are two basic kinds of assets: (1) tangible, which includes all of the physical assets, such as buildings, furniture, machinery, computers, etc., and (2) intangible, which includes non-physical assets such as employee knowledge, credit ratings, patents, trademarks, information databases, etc.

Liabilities—the obligations that the organization has to outside creditors, including landlords, vendors, suppliers, banks, venture capitalists, and others. Liabilities are often broken down into current liabilities that are due now or in the near future and long-term liabilities that are payable over extended periods, such as mortgages.

Equity—the value of the organization to its owners, who may be individuals, investment groups, venture capitalists, and/or stockholders.

For accounting purposes, the assets of the organization are equal to the liabilities plus the equity, or

Assets = Liabilities + Equity

The formula means that our total assets equal our liabilities and equity. While it may seem odd that liabilities and equity are added on one side of the formula, consider that liabilities offset some of our assets. If we have borrowed $50,000 from a bank, we now have assets of $50,000, which we can use to buy equipment or to hire people. We also now owe the bank the $50,000 so we have to put it on the other side of the formula as a liability. We are going to use the money to try to increase our assets. If we spend it and generate a return of $70,000, we have increased our assets by $20,000. On the other side of the formula, we would still show the $50,000 that we owe the bank, but we would also now add $20,000 to our equity because we have increased the value to the shareholders by that amount. Thus, as a result of our profit, both sides of the equation have increased by $20,000.

The Balance Sheet Worksheet gives an example of a balance sheet. This sheet is a simple example for a sole proprietorship. Balance sheets are used to show the overall condition of an organization. Annual reports include balance sheets, although they are usually much more complex that the example shown here. To get a loan from a bank or an investment by a venture capitalist, a firm will have to present a balance sheet (along with many other items).

For a technical communication manager, it is important to note that our discipline is one that is wage intensive. In other words, it does not require a large expenditure of capital to run a technical communication organization; the costs are largely based on the wages we must pay employees. This shows up on the balance sheet in the "Wages Payable" section of liabilities. Hence, we are seen as a liability when upper management looks at the balance sheet. What we must learn to do is to demonstrate where on the other side of the balance sheet, the assets side, we make contributions that offset our costs. This is discussed in more detail in the Value Added section.

Income Statements

An income statement shows, for a given period, how much the organization brought in and how much it spent. The bottom line, then, shows whether the organization had a profit or a loss for that period. Internally, most organizations look at income statements by the month, and they often report the results to the financial markets each quarter, which can have a significant effect on their stock price. They use income statements for these quarterly reports. People in management study them constantly to see where they can make improvements to the bottom line, which, of course, can be done either by increasing revenues or by decreasing costs and expenses.

The Income Statement Worksheet gives an example of an income statement for a sole proprietorship.

Note that the income statement has a list of expense categories. In some organizations, technical communication is included in that list, as if it were equivalent

Technical Communication Management Worksheet

Balance Sheet

Aardvark Development Company

Balance Sheet

Month, Year

Assets			Liabilities and Capital		
Current Assets			*Current Liabilities*		
Cash	$5,700		Accounts Payable	$4,200	
Accounts Receivable	15,100		Wages Payable	16,700	
Prepaid Insurance	1,500		Total Current Liabilities		20,900
Total Current Assets		22,300	*Long-Term Liabilities*		
Fixed Assets			Bank Loan	10,000	
Equipment	32,700		Loan from Principal	20,000	
Less: Accumulated Depreciation	18,400		Total Long-Term Liability		30,000
Total Fixed Assets		51,100	Total Liabilities		50,900
Total Assets		$73,400	*Equity (Capital)*		22,500
			Total Liabilities + Equity		$73,400

Technical Communication Management Worksheet

Income Statement

Aardvark Development Company
Income Statement
For the Year Ended December 31, 200X

Revenue:

Gross Sales		0
Less: Sales Returns and Allowances		0
Net Sales		0

Cost of Goods Sold:

Beginning Inventory	0	
Add: Purchases	0	
Freight-In	0	
Direct Labor	0	
Indirect Expenses	0	
	0	
Less: Ending Inventory	0	
Cost of Goods Sold		0
Gross Profit (Loss)		0

Expenses:

Advertising	0	
Bad Debts	0	
Bank Charges	0	
Charitable Contributions	0	
Computer Network	0	
Commissions	0	
Contract Labor	0	
Credit Card Fees	0	
Delivery Expenses	0	
Depreciation	0	
Dues and Subscriptions	0	
Insurance	0	
Interest	0	
Maintenance	0	
Miscellaneous	0	
Office Expenses	0	
Operating Supplies	0	
Payroll Taxes	0	
Postage	0	
Professional Fees	0	
Property Taxes	0	
Rent	0	
Repairs	0	
Telephone	0	
Travel	0	
Utilities	0	
Wages	0	
Total Expenses		0
Net Operating Income		0

Other Income:

Gain (Loss) on Sale of Assets	0	
Interest Income	0	
Total Other Income		00
Net Income (Loss)	$	0

to buying stamps or paying for overnight delivery charges. As a technical communication manager, you want to ensure that you are included in the same categories as other developers and not considered to be merely another expense.

Note also that when upper management looks at an income statement, they see two main ways to improve their bottom line number: increase revenues or decrease expenses. Even if technical communication does not have its own expense category, it is still thought of as an expense, largely coming under the wages expense category. As with balance sheets, technical communication managers want to look for ways to show how they improve the income statement. They can do so by trying to show how they add to revenues and how they help to reduce expenses. Again, this will be treated in more detail in the section on Value Added.

Budgets

Technical communication managers are more likely to use the department or group budget statement than other financial documents. Learning how to read and understand budgets is an important skill for any manager. Many organizations assume that technical communication managers do not want to receive such documents or that they will be incapable of comprehending them. You should request a monthly breakdown of your group's budget and learn how to read it and use it. This puts you on a par with managers in other departments and it also helps you try to get a fair apportionment of annual and project budget allocations.

Technical Communication Budgets

Your group or department's budget is basically an income statement that includes additional information about the most recent budget period, usually a month, and about budget performance year-to-date. The Technical Communication Budget Worksheet shows the categories from the income statement, with additional columns added for monthly and year-to-date performance.

The budget has eight columns:

1. The first column contains the revenue and expense categories. The categories included on this example are by no means universal, and your budget report may not have the same categories.
2. The second column shows the budgeted amounts for each revenue and expense category for the most recent budget period, in this case a month.
3. The third column shows the actual amounts that were brought in or spent for each category during the month.
4. The fourth column shows the difference, or delta, between the budgeted amount and the actual amount. If you overspent your budget in a category, you will see that amount reported in parentheses (or in red, if your accounting department uses color).
5. The fifth column shows budgeted amounts for the year-to-date. Note that in some organizations the fiscal year is not the calendar year. For various

tax and historical reasons, some companies start their fiscal years on dates other than January 1, usually a date coinciding with one of the year's quarters (April 1, July 1, or October 1). Before you can understand year-to-date figures, you need to know when your organization's fiscal year starts.

6. The sixth column shows actual amounts brought in or spent to this point in the year.

7. The seventh column shows the difference, or delta, between the budgeted amount and the actual amount spent so far this year.

8. The last column, which is not included on all budgets, shows a projection for how you will come out on each budget category if your current trend continues.

As you can see, the budget report gives you a very good idea as to where you stand in each category. You can quickly see those areas where you are spending too much or not enough. You can also bet that your boss and others will be looking at those areas. In most technical communication groups, the manager will have control over only a few of the budget categories. Most of the others will be pre-determined by organizational policies and contracts. Therefore, figures will simply be plugged in to show you how much it costs to run your organization.

Making Cuts to a Technical Communication Budget

By far the biggest expense category on most technical communication budgets is wages. While you may pay allotments for real estate, network services, etc., the only large number over which you have some control is the wage category, which depends, of course, on the number of employees you have and on how much you are paying them. When a technical manager gets pressure to reduce expenses, he/she has only a limited range of possibilities. Yes, if you are in that situation you can try to cut postage costs or office supply costs, but these are usually a small percentage of your wages budget. However, because categories like postage, telephones, and office supplies are the only ones over which many first- and second-level managers have control, we often see the ridiculous situations where squeezed managers issue proclamations that no one can buy another Eagle No. 2 pencil for the remainder of the year.

You can receive that kind of pressure even if you are coming in under your budgeted amounts every month. Note that if insufficient revenue is coming in to pay for all of the expenses, the bottom line is still a negative number.

A technical communication manager under pressure to reduce the budget significantly has only a few options, all of them bad. If the pressure is indeed significant, cutting postage, telephone calls, etc. is not going to create enough savings. You will have to reduce the staffing costs. There are three ways for a technical communication manager to do so:

- Reduce staff and attempt to get the work done with fewer people.
- Lower the average salary you are paying, usually by terminating experienced people and hiring less experienced people.
- Make some staff reductions and hire contractors to do some of the work.

Considering Loaded Rates When Hiring Contractors

An important financial concept for technical communication managers to understand is the "loaded" cost of having an employee on staff. While it is tempting to view the cost for an employee as simply being his/her salary, in reality that cost includes much more. You have two kinds of expenses associated with every employee on staff, direct and indirect costs.

Direct expenses are those "directly" associated with having the employee on staff. These include all benefits such as taxes, health costs, pensions and 401k maintenance with any matching costs you pay, annual bonuses, and other costs. Direct expenses generally cost something like 30 to 60 percent of an employee's base salary, depending on your organization's benefit structure.

Indirect expenses are all of the other costs associated with having the person on staff. You pay for real estate space, for light, for heat and air conditioning, for computer equipment, software, building maintenance, cafeteria subsidies, and many other expenses. In some organizations, the costs for non-revenue generating departments are also added into the indirect expenses. For example, you might be charged an overhead cost to pay for your legal department, whether you use them or not. Note here that you do not ever want the technical communication group to be funded by overhead charges made to all departments. Rather, you want to be charged directly to the projects and departments that use your services.

The total loaded rate for an employee, then, is much higher than just the base salary. It can be expressed in the formula:

Loaded Rate = Salary + Direct Expenses + Indirect Expenses

In most organizations, the actual cost (the loaded rate) for having an employee on staff is 200–400 percent of the employee's base salary. Lean, efficiently managed organizations can get the rate under 200 percent, while bureaucratic, inefficiently managed organizations can run 300–400 percent and even higher.

When you hire a contractor, you generally pay an hourly rate that is much higher than what you have to pay for an employee. However, you do not have to pay the contractor any direct expenses. Those are usually built into the rate charged by the contracting company, along with their own overhead costs plus their profit.

It is important when considering contractors to compare the total cost of having a contractor on staff versus the total cost of having an employee. To do so, you will have to know how your organization handles contractor costs. If you have a contractor working on-site, does your organization load indirect expenses onto the rate paid to the contractor to arrive at the real loaded cost for the contractor? If so, then the contractor may be more expensive than an employee because contractor hourly rates are higher than employee hourly rates. This is a moot point if the contractor is going to perform the work off-site because then indirect expenses should not be applied to what they are charging you. In that case, it will almost always be less expensive to hire the contractor.

Hiring contractors to reduce budget expenses may seem like a tempting solution, but you have to make sure that you are comparing the total costs for a contractor to the total costs for an employee. As discussed in Chapter 2 under assessing

Technical Communication Management Worksheet

Technical Communication Group Monthly Budget

Aardvark Development Company
Monthly Budget Report – Tech. Comm.
For June, 200X

	June Budget	June Actual	June Delta	YTD Budget	YTD Actual	YTD Delta	Proj. for YR
Revenue:							
Documents							
Training Videos							
CDs							
Gross Sales							
Less: Sales Returns and Allowances							
Net Sales							
Expenses:							
Advertising							
Bad Debts							
Bank Charges							
Charitable Contributions							
Computer Network							
Commissions							
Contract Labor							
Credit Card Fees							
Delivery Expenses							
Depreciation							
Dues and Subscriptions							
Insurance							
Interest							
Maintenance							
Miscellaneous							
Office Expenses							
Operating Supplies							
Payroll Taxes							
Postage							
Professional Fees							
Property Taxes							
Rent							
Repairs							
Telephone							
Travel							
Utilities							
Wages							
Total Expenses							

staffing needs, hiring contractors is an excellent solution to occasional workload peaks. However, it may not be the best solution when you are trying to reduce costs.

The Annual Budget Process

Many organizations begin months in advance to prepare their budget for the following year. This process can consume considerable management time and effort. If someone else has been preparing the technical communication budget in your organization, you definitely want to get involved in the process, for several reasons. First, you will have more of a say over how much you get and whether it is adequate to prepare high quality documentation solutions for your customers. Second, you will move toward being regarded and treated like other managers. And third, you can use the budget discussions to educate and lobby other managers about the importance of technical communication to meeting your organization's mission and goals. Budget discussions may not always be pleasant experiences, but they offer an excellent forum for a technical communications manager to use in emphasizing that we understand and contribute the organization's overall mission.

Because annual budget processes are so variable, I am going to discuss only a few general concepts here. For technical communication budgets, the largest expense is nearly always going to be wages, which in budget discussions is often converted into "headcount." A headcount means what it costs to have one person on salary for one year. You can usually convert headcount to dollars by multiplying by some standard number that your organization uses. You also want to know what your organization's assumption is about the number of hours per year that a "headcount" works. While it may seem tempting to use 2,000 hours, which is simply 50 weeks times 40 hours/week (with two weeks taken out for vacation), this number will always be too high. You have to factor in all of the other hours the average employee is not available for charging to a project, including sick time, leaves, training time, internal (non-project) meetings, even the organization picnic. Most organizations use a number around 1,800 to 1,900 hours.

The budget process should involve several steps, as follows (Finkler 1996):

1. **Preliminary planning**. First, you assess the state of your group and any known changes that you will be making during the period for which you are budgeting. Are you currently short-handed or over-staffed? Will you be introducing new developmental tools next year that will require extensive training? Are big projects coming along that will require more people?

2. **Forecasting.** This is the heavy lifting of budget preparation. What are the known projects your group will work on and how many headcount will they require? What new projects do you anticipate, and how many people will they require? You will have to work with the scientists/technical people to find out what they are anticipating for the period so you will have a better idea of the documentation needs for new products and services that are planned. When possible, you want to learn the dollar amounts they plan to spend on those projects, which will help give you an idea of their scope

and of how many technical communicators will be needed. As Hackos (1994) points out, the best way to gather this information is through constant networking within your organization, especially with the marketing and engineering personnel who are involved in planning new projects and initiatives.

3. **Budget Preparation.** Here you convert all of the known and anticipated projects into dollar and/or headcount numbers. You also add any equipment and software expenses you will incur. Obviously, if you are projecting that you will need five more people next year, you will also need five more desks, computers, and software licenses for each program you use. You will also need to increase your numbers for supplies, telephone charges, and all other expenses by the percentage increase that you will experience in your group.

4. **Evaluation/Negotiation.** This is one of the most difficult, yet most important, tasks that a technical communication manager does. Because many scientific and technical personnel do not understand or appreciate the value of properly documenting products and services, they are likely to try to keep technical communication expenses at such a low level that you cannot possibly provide quality products. This part of the budget process is one of the most Machiavellian parts of a manager's job because you are competing for resources with other managers with whom you hope to work amicably for the remainder of the year. Many technical communication managers who have done everything else right lose out during this process, thereby ensuring that they will be understaffed, overworked, and unable to produce reasonable quality with the resources they have. Three things will help you avoid this:

   1. Have the latest and best data about the value that technical communication adds to your organization's products and services. See the following sections of this chapter for more information.

   2. Have a set of quality level definitions that you have converted into service-level agreements. If a projected budget shows you getting less than your lowest-level definition requires, refuse to do the work. If you can't do that, promise that quality or quantity or both will suffer greatly.

   3. Develop an ally in management who will help during budget negotiations. This might be someone in marketing or someone in engineering. Even though everyone tends to look out for their own bailiwicks during budget negotiations, you might be lucky enough to find someone who is aware of the positive impact of quality documentation.

Padding

In some organizations, everyone who is experienced in the budget process knows that upper management will cut their proposed budget by a certain percent every year. So, they simply "pad" it by adding that percent to what they think will actually be required. This is a ridiculous game, but one that is played out in nearly all organizations. If the only way to get the budget you need is to play the game, then do it.

There is a temptation for technical communication managers to differentiate themselves from other managers by showing that we do not play games and that we are more honest and ethical than others. So, we will show them all by turning in a budget that is exactly what we need. When everyone else ends up with the number they wanted (after their padding was cut), and we end up with 20 percent less than we needed, we are left on the moral high ground and the fiscal low ground. We are also left facing a very difficult year trying to meet project obligations without the resources to do so.

Levels of Quality and Service-Level Agreements

Many managers in scientific and technological organizations do not place a high value on technical communication. They have received little or no education about documentation and its value. Worse, they may have received instruction telling them that product documentation should cover every detail of how a product works. Programmers, for example, are taught to "document" their code, meaning that they put a comment on every line explaining what the function of that line is. In most engineering and programming courses, customer documentation is mentioned as an afterthought, if at all.

Getting product and project managers to commit reasonable funds for technical communication products can, therefore, be quite a challenge. This is especially true in an environment where everyone believes that their efforts to create quality products are compromised by lack of enough funding.

Levels of Quality

One of the best ways for technical communication managers to educate other managers about the value of good documents and the necessity for adequately funding documentation is to develop definitions for levels of quality. To do so, the technical communication manager develops several quality levels along with the criteria for each one (See 1995). Table 3.1 contains an example of one possible system of quality levels.

The purpose of the quality levels is to show other managers which of the steps in your quality process you cannot do without sufficient funding and to show them all of the steps involved in producing high quality technical communication products. When confronted with the reality of what you will omit if they opt for a lower level, they may choose a higher one than they otherwise would. In any case, the levels help you manage their expectations about what they will get for a certain amount of funding.

In some organizations, you can tie levels of funding to each quality level based on percentages of the overall project budget. This is a potentially dangerous practice, but if you have some historical data, your estimate of budget requirements will

TABLE 3.1	Technical Communication Quality Level Example											
Quality Level	Criteria	Front-end analysis	Customer analysis	Task analysis	TC Plan	Content Spec.	Optimize design	Write & revise	# of reviews	Usability testing	Substantive Edit	Index
1 - Basic	Includes only revisions to show updates and new features, no design or improvement							Yes	2			
2 – Good	Includes standard quality steps		Yes		Yes	Yes		Yes	2		Yes	
3 – Best	Includes all quality steps aimed at best possible documentation & training solutions for the customer	Yes	Yes	Yes	Yes	Yes	Yes	Yes	2 or more	Yes	Yes	Yes

usually be close. For traditional software documentation projects, for example, something approximating the following levels works most of the time:

Level 1 – Basic   5-10%
Level 2 – Good   8-12%
Level 3 – Best   12-15%

The problem with such percentages is that some projects can vary radically, depending on their nature. A Web site, for example, might require that nearly 100 percent of the funds go to the technical communication group, assuming that the Web developers are in that group. On some software projects, the size of the online help and tutorial code is greater than the product code, so getting only a small percentage of the overall budget does not provide enough funding.

To prepare your own levels of quality, you will have to decide which technical communication quality steps are most significant in your organization and then build a table similar to the one above. The Technical Communication Levels of Quality Form will help you do so.

Another thing the levels of quality will help you do is to say no. If an internal group is trying to fund you at a completely unreasonable level, you can use the table to defend your refusal to do the work. Over time, you should develop historical data for every project that your group works on and the level of quality at which you performed the work. If your organization gets customer feedback or

Technical Communication Management Worksheet

Technical Communication Levels of Quality

Quality Level	Criteria	Front-end Analysis	Customer Analysis	Task Analysis	TC Plan	Content Specification	Optimize Design	Write & Revise	# of reviews	Usability Testing	Substantive Edit	Index		
1														
2														
3														
4														
5														

To create your own levels of quality, decide how many levels you want and what you want to call each one. Of course, your naming techniques will have some rhetorical implications. Decide the basic criteria for each level, and then check the quality steps that will be included with that level. You can add as many additional steps as you need to.

The table will not only help you conceptualize how the interplay between quality and funding works, but it will also help other managers see those dynamics, too.

satisfaction measurements of some kind, you can compare those results to the level of technical communication quality that was used for each project. You should see a correlation between how satisfied your customers are and how much quality effort went into preparing the technical communication part of the product. You will then have even stronger data to show in support of adequate technical communication funding.

Service Level Agreements

Managers in computing and other service-oriented industries are familiar with service level agreements (SLA). An SLA is basically a contract between a service provider and a client that specifies the level of service that the client will receive for the price being paid. Many organizations, for example, hire outside firms to provide computer network services. The outside firm may offer service levels ranging from off-site administration with no guarantees about network down time, all the way to on-site management with guarantees of 100 percent up time. The vendor and the customer negotiate the level of support needed and then sign an SLA that specifies everything in considerable detail.

A technical communication manager can use service level agreements to specify how much work and what level of quality an internal or external client will receive for the amount of funding they are willing to commit to a technical communication project. Using SLAs or other contracts helps spell out for everyone involved how much effort communicators will expend on a project, what the assumptions are about the scope and length of the project, how many communicators will work on the project for what durations, and other specifications. This information is often spelled out in information plans (or documentation plans, technical communication project plans, or whatever you call them in your organization). If it is, and if these plans are reviewed and signed off by all project groups, then you do not need to worry about SLAs. However, if you do not have a formal information plan signoff structure, you may want to use an SLA or a contract of some kind.

One of the biggest problems in managing technical communications functions comes from the fact that managers in other areas who retain the services of technical communicators have no idea what they do or how much it should cost. Consequently, they grossly underestimate the cost for documentation. Depending on the structure of the organization, the technical communication group may be forced to do the work regardless of how penurious the budget is. The result is poor quality documents, unhappy external and internal clients, overworked and stressed communicators, and dissatisfied managers both inside and outside of the technical communication group. Using an SLA or a contract can help avoid this all too common scenario.

While SLAs are usually applied to computing and telecommunication services, a technical communication manager can translate the concept to our discipline fairly easily. For more extended definitions and examples, simply search for "service level agreement" on the Internet.

Contracts

Technical communication groups often share a kind of victim mentality. They perceive that they are second-class citizens who are underpaid and under-appreciated in the scientific and technological environments in which they work. Engineers and programmers get all the glory and the money while the communicators work seventy-hour weeks to compensate for inadequate funding and uncontrolled project changes. While this is true in many organizations, communicators share some of the blame themselves.

Technical communication managers need to learn the same business skills that other managers have. This includes doing work on a contractual basis, even internally, rather than simply agreeing to everything, no matter how impossible and debilitating it is going to be. One of the best ways to begin to get respect within scientific and technological organizations is to contract for work to be done. Insisting on clear, well-defined deliverables, due dates, and levels of quality will move a technical communication organization toward being regarded as more professional and reliable. Even in what Hackos (1994) calls a "clueless" organization, perhaps especially in such an environment, communication managers need to develop contracts that define work commitments before they are done.

A good technical communication contract has, at a minimum, the following sections:

1. A statement of the goals of the technical communication commitment. This can be very specific when the project is already well-defined, but at other times it will have to be more general to allow for what is discovered during the analysis and planning stages of the technical communication development effort.

2. A statement about final deliverables that is as specific as possible, given what is known so far about the product or process being documented.

3. A list of key milestone dates that includes everything you will deliver and every date by which the client is responsible for providing information and returning review drafts.

4. A definition of each deliverable, spelling out what it will contain and how complete it will be. For example, a deliverable definition for a first draft might specify that at least 75 percent of the technical information will be included, that 50 percent of the graphics will be included with the other 50 percent as hand-drawn sketches, that it will not have an index, etc.

5. A list of the personnel you will assign to the project and the role/responsibility of each.

6. A list of the client's personnel who will be responsible for providing information, reviewing drafts, signing off on final versions, and other functions.

7. An estimate of document sizes and hours to be worked on the project by each of your personnel, converted to dollars.

8. A set of assumptions on which the contract is based, requiring that the contract be altered if the assumptions change. This is especially important on projects where "scope creep" is anticipated. Further, the contract should

include specific procedures for increasing the estimate dollars when the scope does increase.

9. A clear, definite statement as to whether the project is being done on a fixed price or an hourly rate basis. (See Estimation for more.)

10. Identification of one prime client contact who is responsible for resolving contradictions, choosing among conflicting priorities, and assisting with client personnel who are uncooperative and/or who miss deadlines.

11. A statement about overtime policy stating whether the client pays at the regular hourly rates or at higher rates.

12. The usual contract language concerning how and when payments will be made, who can terminate the contract and why, etc.

13. Signatures with dates for the key parties involved.

While it is beyond the scope of this book to offer an extended tutorial on how to write a contract, the Technical Communication Contract form will get you started. You can also consult with Hackos's (1994) Chapter 8 on Project Plans, which are similar to contracts. Further, see Frick (2001).

Estimation

All technical communication managers will be asked to supply estimates for overall projects and for individual documents. Knowing how to estimate development costs accurately is an especially valuable skill, whether one is doing so for one's employer or for one's own company or contracting effort (when it becomes even more valuable). Even on projects that will be done on an hourly basis rather than a fixed cost basis, clients (internal or external) will want an estimate as to the cost and time required to prepare the technical communication components of the project.

To the newer technical communication manager, estimating can seem like a daunting task shrouded in mystery, with so many variables that any hope of getting close is based more on luck than on technique. However, following some general precepts, along with some more specific steps, can lead to generally accurate project estimates. This section will first discuss the general precepts and then the step-by-step procedure.

Estimating Precepts

An estimate is an estimate is an estimate. That is, it is an educated guess, not a contract. At some point it may become a contract, but that should not happen until all possible background information has been gathered and all possible variables have been considered and factored in. Managers of scientific and technological groups often assume that we can estimate accurately off the top of our heads, and that we should be able to give them a firm figure with only a minimum of background information about the nature of the project. They are also often stunned by the cost of creating high quality technical communication products. Remember, though, that an estimate is an approximation, and that all of the

Technical Communication Management Worksheet

Technical Communication Contracts

Technical Communication Goals: Solve what problem, document what product, etc.

Final Deliverables: Deliverable title, size (pages, screens, etc.)

Milestone Dates: Tech Comm. Org. — specific personnel

Client Org. — specific personnel

Deliverable Definitions: Information Plan

Content Specification

First Draft

Second Draft

Final

Tech. Comm. Personnel: [Name, project role]

Client Personnel: [Name, project role]

Estimate: [Product size, hours, dollars]

Assumptions: Scope, information availability, etc.

Method for cost adjustments

Pricing Basis: Fixed rate or hourly

Client Primary Contact: One specific person "in charge"

Overtime Policy: Regular rate, increased rate, maximum

Other Standard Contract Sections: Payment schedule and method

Termination policy

Liability, etc.

Signatures: Tech. Comm. Manager, principles

Client manager, principles

assumptions associated with arriving at it should be communicated along with the bottom-line amount.

The first number you say aloud or put in writing is the number people will remember. A corollary is that if you give a range, the lowest number is the one that people will remember. You get stopped in the hall some evening at 6:00 P.M. by a project manager who asks for a "ballpark" price for an on-line help system for his new project to begin in a couple of months. He says he thinks the project will be similar to the ABC project, where the documents cost $50,000, so he wants to use that number. Sure, you say, as you head for home. Later, when you gather the specifications, requirements, and other information about his project, you realize that the cost is going to be more like $100,000. When you turn in the estimate, he goes ballistic, accusing you of increasing *your* original estimate by $50,000. The best policy, then, is simply to refuse to give a number until you can get enough project information to allow you to make a reasonably accurate estimate.

Those who do not learn from the mistakes of the past are doomed to repeat them. The best way to do estimates is based on sound historical data. A thorough tracking system for each technical communication project can provide information that takes into account the special variables in that organization that affect the costs of creating technical communication projects. In the absence of a tracking system, you can use the experience of earlier projects for similar types of projects. If an earlier project of similar size and scope in your organization cost a certain amount, you can at least have some confidence in a similar estimate for your upcoming project. However, this can be very misleading, especially if you are documenting new technologies or using new tools or media yourself.

With apologies to John Keats, heard melodies are dear, but those unheard are dearer. With estimates, it is important to consider every assumption you have made. The implicit assumptions that you do not consider and do not state are the ones that can cost you dearly. This is particularly true for contractual work, where not stating assumptions clearly and comprehensively can lead to huge overruns with no way to recover, so that you lose money on the project.

The later you make an estimate, the more accurate it will be. Boehm (1982), for example, has shown graphically that estimates done during a project's feasibility stage can be off by as much as 400 percent, while those done after detailed design specifications are published are generally within 15 percent. Only on poorly run projects (and there are many of them) are teams required to supply early estimates and then try to live with them, even when additional information makes it totally clear that those estimates are off. Estimation should be an ongoing part of a project, with reassessments and adjustments made continually.

There is a correlation between quality and funding. Like it or not, all technical communication managers must spend considerable time struggling with the eternal and inevitable tradeoffs between desired quality and actual funding. Polson (1988) provides the following formula for seeing the relationship between quality and funding:

Function x Quality = Resources x Time

As the formula shows, an increase in product functions or quality requires an increase in either resources or time (or both). If resources are constrained, then either function or quality must be reduced.

This reality creates one of the technical communication manager's most burdensome tasks. No one wants to tell communicators that they should do less than their best work; however, technical communication managers, faced with the realities of the funding amounts they get for projects, must often do so. This is the reason that the concepts of quality levels and service level agreements are so important (see earlier sections in this chapter). All technical communication managers must decide how far they will go in reducing quality in order to meet budget realities, because they will be pushed to that point by scientific and technological colleagues who do not believe that any money at all should be spent for documentation. Each of us must decide when our quality standards are so compromised that we refuse to do the work and simply walk away from a project or from an entire organization.

The estimation process should not obscure the front-end assessment process. It is all too easy to respond for a request for an estimate on a 300-page user's guide by simply doing some math and coming up with an answer. However, a technical communication manager should first look at the larger picture and ask whether a 300-page user's guide is the right solution for a particular project. Technical and scientific project managers often simply assume that a specific solution is the right one, either because they cannot conceive of anything else or because "that's the way we've always done it," despite evidence that suggests no one even looks at their massive user tomes (Mirel 1993). Instead, a technical communication manager should first request the opportunity to perform an audience analysis and a user-task analysis, perhaps involving field studies, before arriving at the documentation deliverables for the project.

Account for all of the variables that can affect your estimate, or they will bite you. Many variables affect documentation projects. The project, through no fault of our own, slips its deadlines repeatedly as product developers go through unanticipated re-designs. A key technical communication person is out sick for three weeks. We cannot get any information from the SMEs, who are swamped themselves. We are spending way too much time learning a new software tool that we have to use to create the graphics. When doing an estimate, you must take as many variables as possible into consideration. Hackos (1994) suggests several standard technical communication variables (or dependencies, as she labels them), in Table 3.2.

Hackos assigns a range of plus or minus ten percent to each of these variables. In other words, if we have excellent SME availability, it will reduce our project cost from 100 percent down to 90 percent, so we save 10% because the SMEs are readily available. Conversely, if we have little or no access to the SMEs, it increases our project cost by as much as 10 percent, so that this one variable could make the project cost 110 percent of what it otherwise would. My own experience indicates that the following additional variables should be considered, with ranges indicated in Table 3.3.

For your estimating purposes, you will have to decide which of the variables listed in these two tables affects your projects and how much of an effect it has. You might be able to identify the ones that apply most aptly to your situation

TABLE 3.2	Hackos's Variables for Estimation
Variable	**Range**
Product Stability	90–110%
Information Availability	90–110%
Subject Matter Experts	90–110%
Review	90–110%
Writing Experience	90–110%
Technical Experience	90–110%
Audience Awareness	90–110%
Team Experience	90–110%
Prototype Availability	90–110%

TABLE 3.3	Additional Variables for Estimation
Quality Process—Internal	80–120%
Quality Process—Client-dictated	70–130%
Technical communication tool experience	90–110%
Project team maturity	90–110%
Equipment/network problems	80–120%

and to develop a standard set of variables to use in estimating. To determine what number you use for a particular variable, consider whether this variable is likely to be better than average, average, or worse than average. If you think, for example, that you will have ready access to SMEs, you might want to set the SME variable at its lowest point, 90 percent, which will reduce your estimate amount. On the other hand, if you know that access to SMEs will be very difficult, you can set the variable at its highest point, 110 percent. If you believe that you will have a mixture of SME cooperation, you can set the number at 100 percent, which will not affect the final estimate. It is a good idea to track such variables over time and to learn which ones to use for your organization and where to set them for the various project teams with whom you work.

Also, you should consider additional variables that are specific to your organization and its processes. Anything that can increase or decrease a technical communication project's cost should be included. As the extended budget example below demonstrates, the original development estimate can easily go up by a factor of 50 percent when all of the variables and other costs are added.

Fixed Price versus Hourly Charges

Some customers, whether they be internal or external, want a fixed cost estimate for how much they will have to pay for a set of documents. Others are willing to pay technical communication costs on an hourly basis. They may or may not have asked for an estimate of the total number of hours (and hence cost) initially. Often, they will want a "not-to-exceed" amount (Zvalo 1999).

It is easier for a client to administer a fixed cost agreement. The client also does not have to worry about the efficiency with which the communication group is working. The onus is on the communication group to get the work done within the fixed cost amount while maintaining the agreed-upon quality. Hence, on a fixed cost contract a technical communication manager must carefully track costs and progress.

Because there are so many variables associated with preparing documentation, many of them outside of the technical communication group's control, charging hourly rates is always a safer approach for the communication group. If a client insists on a fixed cost, it becomes crucial for the communication manager to spell out carefully all of the assumptions behind being able to meet that cost, especially those regarding the scope of work and the number and nature of changes that the client can make to the original product definition. Without carefully defining the scope and how changes will be agreed upon, the technical communication manager risks that the project will, as do many, grow in scope well beyond the original conception. Without some means of redefining the scope (and hence the price), the communication manager is left with the choice between completing the work no matter how much time and effort it takes (and hence losing money) and completing the work within budget but doing such a shoddy job that document quality is greatly diminished.

For either type of estimate, the technical communication manager should go through the entire process described in the next section. Failure to do so can lead to estimates that cause a loss of revenue or of reputation.

The Estimating Process

Following a repeatable estimating process provides the best chance for making reliable estimates. It is tempting, especially for experienced managers, to simply lick one's finger, hold it up the wind, and proffer the first number that comes to mind. Those with extensive experience at estimating can often come very close without following any formal methods. For less experienced managers, however, following a set procedure can help reduce the chances for wide variations. First, let's look at a formula designed to yield technical communication estimates.

TPC = (((Units x hours/unit x cost/hour) x variables) x markup) + indirect costs

Explanation:

TPC = Total Project Cost.

Units = the types of technical communication deliverables you are creating and the smallest segments into which you can divide them (Smith, 1994). For example, the obvious unit for a book is the page, for an on-line help system is the help screen, for a training video is the minute, and so on.

TABLE 3.4	Industry Averages for Technical Communication Unit
Unit	Hours to Develop Each Unit
Software documents	4–5 hours per page
Hardware maintenance, theory of op.	8 hours per page
Classroom training	20–40 hours per class hour
On-line help	4 hours per topic
Video	30 hours per minute
CBT	60 hours per finished hour
Light editing	4–8 pages per hour
Substantive editing	2–4 pages per hour

Hours/unit = the number of hours required to develop one unit. Ideally, these are based on historical numbers in your organization, but they can also be based on industry averages (Lang and Ricks 1994; Hackos 1994). Note in Table 3.4 that these hours include all technical communicator time from the beginning to the end of the development project. They include much more than the actual time spent writing, which is usually a small percentage of the overall time.

Cost/hour = your total "loaded" cost for one hour of technical communicator time. This can be based on the actual loaded salaries of the communicators involved or on some average number developed for your technical communication group or organization.

Variables = all of the factors that can influence whether a project costs more or less than "normal." Such factors will vary among organizations, but universal variables in technical communication include access to information, SME availability, number of review cycles required, experience of the communication team, and others. For many of these variables, the impact on a project has about a plus or minus 10 percent effect. Others might swing as much as 20 to 30 percent. For example, if the project requires that you follow a stringent set of quality procedures, it might increase the cost by 20 percent. You would therefore need to multiply the project cost times 1.2 (or 120 percent) to arrive at the adjusted cost. You could multiply the cost by each variable, one at a time, to get the overall impact of the variables. However, it is easier simply to multiply the variables by each other to arrive at a variable factor. You then multiply that factor times the cost so that you increase or decrease the cost by the variable factor.

Markup = the percentage by which your organization marks up the base costs for a project. Depending on how your accounting is done, this can include other costs that don't show up in loaded salaries or indirect expenses, or it can include only profit.

Indirect costs = all of the non-salary-related expenses for the project, including, for example, travel, software and hardware purchases, sub-contracted work, office supplies, etc.

Using the formula as a guide, we can go through the procedure for doing an estimate, with the following steps:

1. Gather all information possible about the proposed project, including audience and task analyses, marketing analyses of the audience, records of similar products or earlier releases of this product from customer support, engineering specifications, programming requirements, etc. Make your estimate using as much relevant information as possible.

2. For each type of deliverable, estimate the number of units that will be required to meet your audience's needs. Some organizations might be able to arrive at guidelines for this calculation. For example, you might know that it takes ten pages, on the average, to document each new feature for a piece of software, or ten pages for each new procedure you add to a set of practices and procedures. If you know how many new features or procedures you have, you can simply multiply that number by the average number of pages required to produce each. This is also a good time to look at previous experience on similar types of projects with similar products and procedures. If your organization does not have prior experience on the type of product or process you are estimating, you can look at the size and scope of documents that competitors have used (although you might want to try to do better).

3. Calculate how many hours it takes to complete the average unit. You should have this from your own historical data from previous projects. If not, you can use the industry averages included in Table 3.4, but beware that those averages may not apply to the particular development environment in which you are working.

4. Multiply the number of units by the number of hours needed to complete each unit to arrive at the total hours required to develop the deliverable.

5. Multiply the total hours by your cost per hour. As noted above, this might be an average supplied by your accounting department or it might be based on the loaded rate for the employee who is actually going to do the work.

6. Based on what you have learned about the product and the project team that will work on it, assign to each of your variables a number within its normal range. For example, assume that you have the following five variables with the ranges indicated in Table 3.5:

TABLE 3.5	Variable Levels for a Sample Project	
Variable	Range	This Project
SME Availability	90–110%	90%
Product Stability	90–110%	110%
Team Experience	80–120%	100%
Review Cycles	90–110%	110%
Quality Process	80–120%	120%

For this case, you have decided that you will have ready access to SMEs, and hence you have used the lowest number in the SME range. Notice that this will reduce the project cost by 10 percent. For product stability, because you are working on a brand new product, you have chosen the highest number in the range, which will increase the documentation cost by 110 percent For team experience, you have decided that your team has worked together before, but not frequently, so this is neither a big advantage nor disadvantage, and you have chosen 100 percent, which will not change the project cost. For review cycles, you know that with a new project and with marketing's plans to have several trials with customers you will be asked to produce numerous review copies and interim drafts. So, you have assigned review cycles to the highest number in the range. Likewise, you will be required to use your organization's most stringent quality process, which adds some development steps and some accounting time, so you have set that number at 120 percent. To arrive at your overall variable factor, you can now simply multiply the numbers in the third column, first converting them to decimals—0.9 times 1.10 times 1.00 times 1.10 times 1.20 equals 1.3068. Consequently, you will have to multiply your project cost times 1.3 to account for the effects of the variables on your documentation development cost.

7. You now multiply the total technical communication development cost by any markup that your organization requires. You will need to get details about this from your supervisor or your accounting department. In some cases this number is built in to the hourly loaded rate you use. If so, you omit this step. If not, you simply multiply the cost you arrived at in Step Six by the markup percentage and then add the resulting number to the total project cost.

8. Indirect project expenses are usually not included in the markup multiplication but are added on. In other words, your organization is not making any profit on the expense items. However, some organizations do add some of the expense items into the project amount before they apply the markup. You will need to ask your supervisor or accountants how they handle this. Otherwise, add the direct costs you expect to incur to the total project cost calculated in Step Seven to arrive at your final estimate.

To better illustrate how this works, we will follow an extended example through the estimation process. We have been asked to develop a 300-page user guide. However, after our audience and task analyses, we conclude that we will instead do a 50-page "getting started" guide and an on-line help system. To calculate the cost to develop the guide, we go through the 8-step process.

1. We gather all pertinent information about the project and variables that will affect our development of the guide.

2. We have already estimated the number of units, in this case pages, at 50.

3. We know from historical data that in our organization it requires four hours, start to finish, to complete a page of a getting-started guide.

4. We multiply 50 pages times 4 hours/page to get 200 hours required to develop the guide.

5. Our supervisor has told us to use an average loaded rate of $70 per hour for a technical communicator. We multiply 200 hours by $70/hour to arrive at $14,000 as the initial development cost for the guide.

6. We now decide on our variables. For this example, we will use the same variables and numbers as listed in Table 3.5. We multiply our development cost, $14,000, by 1.3 to get a revised development cost of $18,200.

7. Our supervisor has told us to use 30 percent as a markup number. We multiply our revised development cost, $18,200, by .3 to get a markup of $5,460. We add $18,200 to $5,460 to get $23,660.

8. We estimate that we will incur the following expenses to prepare the guide:

Travel—3 trips for user testing @ $1,000/trip = $3,000
Software—special graphics product required = $1,000
Total $4,000

We add to $23,660 the expenses of $4,000 to get a total estimated cost of $27,660 to develop the getting-started guide.

You can see that the original estimate of $14,000 has nearly doubled with all of the other factors taken into account. This example demonstrates the danger of merely doing quick estimates without accounting for variables, markups, and expenses. While your organization may require some additional steps to the estimation process, you can use the steps described here to get started doing estimates in a formal, repeatable manner. For your estimating to get better, it is important to keep track of the actual project costs, to compare the actual costs to your estimated cost when the project ends, and to make adjustments so that next time you perform an estimate it will be more accurate.

The Technical Communication Estimation Worksheet should prove to be useful in performing your own project estimates.

Percent of Overall Budget Estimates

Another method for estimating TC projects is to apportion a percentage of the total project budget to documentation or to apportion a percentage of the total staff assigned to the project to documentation (Thomson 1998). Some organizations that keep historical records of the cost for each part of a project can calculate what each function's percentage has been historically. Even if your organization does not keep such records, it is a good idea to calculate them yourself. At one company where such records were kept, it was discovered that there was a correlation between the customer satisfaction ratings that documentation and training received and the percent of the project budget that was apportioned to them. In general, percentages spent on documentation and training that were less than 10 percent of the overall budget led to less than satisfactory customer ratings. That data was used to persuade project managers to devote more resources to documentation. It was also used to establish a baseline percentage of project budget below which the documentation organization simply refused to work. One industry study found an average ratio of one communicator to every eleven developers, not very high given the example above (Barr and Rosenbaum 1990).

The biggest problem with percent of budget or percent of staff estimating is that they do not account for projects where some of the variables are outside of the usual organization practices. A new product type may require considerably more resources for documentation than previous "standard" efforts have required. Thus, while percent of budget or staff estimating can provide an approximate budget amount, it should not replace the careful analysis of variables discussed earlier.

Value Added for Technical Communication

Technical communication managers need to understand the overall goals of the organizations in which they work. They also need to actively contribute to meeting those goals and to demonstrate how the technical communication group contributes to doing so. Many technical communication groups find themselves marginalized and unappreciated in the scientific and technological environments in which they function. The management in such organizations, usually pulled from the ranks of scientists and engineers, frequently does not understand or appreciate the contributions made by technical communicators. They may see the function as a "necessary evil," an expense that has to be incurred because they have to send documents about their products and services.

It is the duty of a technical communication manager to educate upper management about the contributions that communicators make, and to show how communicators add value to the products and services that the organization generates. Bryan (1994), Hackos (1994), and Plung (1994) all discuss the need for technical communicators to prove their value by integrating the communication function to the larger goals and operations of the organization, and to contribute in ways that go beyond merely writing support documents. Failure to do so ensures that the technical communication function will continue to be regarded as an expense item, one that could possibly be outsourced. Perhaps the main argument for how communicators contribute is that they are not merely adding value, but rather their outputs are an intrinsic part of the organization's products themselves. They are not a service that adds value; they are product developers that participate in creating the value that the organization's products and services offer to customers.

Many organizations are becoming more aware that the information they have is their asset of greatest value. Indeed, those organizations that exist only on the Internet are, in essence, products of the communicators who develop and maintain their Web sites. Fortunately, this emphasis on the importance of information has helped the status and the viability of technical communication groups. However, it is still necessary for communication managers to show how their work developing and maintaining the organization's information contributes to its overall goals.

The following sections begin with a formula for calculating how technical communicators contribute value. Subsections then cover how we add value in financial ways and then how we do so in non-financial ways.

A Formula for How Technical Communicators Add Value

It is convenient to create a mathematical formula to look at how we add value. For the part of our contribution that is fiscally measurable, using the formula helps

Technical Communication Management Worksheet

Technical Communication Estimation

Project Name: _____

Estimate Date: _____

Estimated By: _____

 1. Technical Communication Deliverable Type: _____

 2. Number of units (pages, screens, messages, etc.): _____

 3. Hours per unit to develop this type: _____ hours per unit

 4. Multiply line 1 by line 2 and enter result here: _____ total hours

 5. Enter the hourly cost to develop this type of unit: _____ $ per hour

 6. Multiply line 4 times by line 5 and enter result here: _____ $

 7. Enter variables, ranges, and range for this project:

Variable	Variable Range	This Project

 8. Multiply "This Project" numbers times by each other and enter result: _____

 9. Multiply line 6 by line 8 and enter the result here: _____ $

10. Enter your markup percentage here: _____

11. Multiply line 9 by line 10 and enter the result here: _____ markup amount

12. Add line 9 and line 11 and enter the result here: _____ direct costs plus markup

13. Enter estimated expenses (indirect costs) here: _____ expense

Add lines 12 and 13 for Total TC Development Cost = _____

us to consider all of the possible contributions we make. It also helps us show others what those contributions are. The following formula can be used to calculate value added by technical communicators:

$$VA = ((CR+CA+RE) - DC) + IC$$

Where:

VA = Value Added

CR = Cost Reductions

CA = Cost Avoidances

RE = Revenue Enhancements

DC = Development Costs (for technical communication)

IC = Intangible Contributions

Whether we like it or not, upper management tends to define value as the contribution we make to the "bottom line," meaning the financial well-being of the organization. While communicators contribute value in important, non-financial ways, the formula helps us see how we contribute in measurable, financial terms—what Hart (2001) calls "quantitative measures."

As discussed earlier in this section, an organization's income statement provides a picture of its overall financial status. We can improve the bottom-line on the income statement in two primary ways: (1) by reducing the numbers on the cost and expenses side and (2) by increasing the numbers on the revenue side. The formula includes the three types of added value that Mead (1998) defines: reduced investment on communication, improved return on investment, and reduced after sales costs. The formula also includes a slot for intangible contributions. Let us examine each part of the formula in more detail.

Note that the development costs should include hidden costs, such as reviewers' time to examine documents or graphic artists' time to prepare illustrations. These expenses may not be included in the technical communication budget, but to calculate a fair value-added amount, they should be added to the development costs.

Cost Reductions Technical Communicators Can Make

We can look at reducing costs in two ways. The first is reducing costs that we are experiencing now. The second is to avoid costs that we will incur if we do not do anything to avoid them. Cost reduction, covered in this subsection, is often easy to identify and to measure. Cost avoidance, covered in the next subsection, is more difficult to quantify.

A group within a larger organization can consider cost reduction in two ways: internally and externally. Internally, how can we reduce costs within the technical communication group? How can we create documents of the same or higher quality while reducing the development costs we incur to do so? This is usually the first place people look when they are asked to quantify the contribution they make and to improve it. For a technical communication group, it is very tempting to show cost reductions by doing things that may reduce the quality of our products. For example, if we do away with general editing, we reduce our development costs because we no longer have to pay for editing services. The first year we do this, it will look like we indeed reduced costs. However, when the customer feedback starts to show that our documents have more errors and are not as useful as they were before, we may find that we have in fact increased the costs. Another temptation, to which many technical groups are succumbing, is to put everything on-line whether or not that is the most effective and appropriate way to communicate the information to the audience. We can show that we have reduced printing, production, and distribution costs, sometimes dramatically, and we get a few months grace period before customers start complaining that they cannot find the information they need. It is important, then, to look at the effect that cost reductions will have on quality and to count only those that have a neutral or positive impact on quality.

The following table covers ways that technical communicators can reduce costs and methods for calculating how much is saved.

Table 3.6 shows two kinds of cost reductions, process-related and product-related. The process-related cost reductions require finding more efficient ways to do things during development. In the short term we can often uncover inefficiencies in the processes we use. This kind of cost reduction is a dual-edged sword, however. First, we are asking for credit for changing our own flawed ways. Second, there are only so many internal efficiencies we are going to find. While technical communica-

TABLE 3.6	Calculating Value Added Through Cost Reduction Internally
Cost Reduction	**Calculation**
Reduce development costs	Compare current cost per unit to historical numbers (last year it took 4.2 hours to produce a page; this year it is taking only 4.0). Multiply units times the old and the new. For a 100-page document, we would have spent 420 hours, but we spent only 400. The 20-hour reduction times our loaded hourly rate (say $70) equals a $1,400 saving.
Reduce production costs	Compare former costs to current ones. Last year we spent $100,000 on printing and shipping paper documents. This year we spent $10,000 putting everything on CD-ROMs shipped with the product. CR = $90,000
Reduce customer support hotline calls	Compare the previous number of monthly calls to the current number. After we revised the documents for higher usability, hotline calls went from 500 per month to 300. Multiplying the reduction in calls (200) times the cost per call (say $28) equals $5,600. Multiply the monthly amount ($5,600) times 12 for an annualized cost reduction of $67,200.
Reduce required field maintenance trips under warranty	Multiply the previous number of trips times the cost per trip and compare to the current number of trips. Last year we required 48 trips at an average of $2,000 per trip. After revising the installation, administration, and maintenance manuals for greater usability, the trips went down to 26 this year. Multiply the 22 trip reduction times $2,000 for a cost reduction of $44,000.

tors should constantly look for ways to improve the processes they use, trying to find significant cost reductions will eventually reach the point of diminishing returns. Also, upper management is happy to see this kind of cost reduction, but they are more impressed with revenue enhancement and with product-related reductions.

The second type of internal cost reduction, product-related, happens after the product has been delivered to its audience. It reduces costs that we previously incurred because what we delivered was not as easy to use as it is now. Again, we have a dual-edged sword, as we are asking for credit for fixing flaws that arguably should have been avoided in the first place. Nonetheless, the result is a genuine contribution to the organization's bottom line.

While reducing external costs may not contribute directly to our own organization's bottom line, it can be so important for customer satisfaction that it affects whether or not our products and services sell in the first place. We can measure external cost reductions in the same way we measure internal cost reductions, except that in this case we calculate the costs that our customers save (Carliner 1997). Table 3.7 shows some examples.

A final problem with using cost reduction as a way of showing value added is that it provides only short-term credit, typically one year or less. If we indeed reduced hotline calls by $67,200 last year, then next year the baseline number will be $67.200 lower and we will have to look for ways to reduce an already reduced number. It will get harder and harder to rely on cost reduction as a method for showing value added.

Cost Avoidance Technical Communicators Can Contribute

Some cost avoidances are easy to show and to quantify. Others, however, are nearly impossible to prove or to put a number on. For example, it is easy to show that our new installation manual is so good and so clear that we do not have to send an installation team to each customer site, which saves us $5,000 per sale. It is more difficult, though, to prove that our improved maintenance manual will reduce repair costs by a certain amount. Measuring many kinds of cost avoidance, then, can become a murky, nebulous effort. The most successful measures of cost avoidance come when we have historical comparisons documenting improvements we have made. For example, if our software products have averaged a 20 percent return rate and, after making significant improvements to the documents/help systems to improve usability, the new return rate falls to 10 percent, we can claim the 10 percent avoidance in return write-offs. While this is similar to the cost reduction model, in this case we are talking about a product that has never shipped, so we are avoiding potential costs rather than reducing existing costs. Table 3.7 includes examples of cost avoidances.

Revenue Enhancements Technical Communicators Can Make

In many organizations technical communicators do not appear to be in a position to enhance revenue in any direct, obvious way. In others, the technical communication group is considered a "profit center," one responsible for bringing in more

TABLE 3.7	Calculating Value Added Through Cost Reduction Externally
Cost Reduction	**Calculation**
Reduce training costs	Compare new training costs versus prior costs. Subtract the new costs from the prior costs. For example, the old version or a competitor's version required one week of classroom training for 1,000 employees at a cost of $3,000 per employee ($2,000 in loaded salary plus travel and lodging expenses) for a total of $3,000,000. The new training is offered in a tutorial included on our CD-ROM, requiring only four hours of employee time, at $50 per hour or $200 times 1,000 employees = $200,000. The old training cost of $3,000,000 minus new training cost of $200,000 yields a $2,800,000 cost reduction.
Increase time to productivity	Compare old time to productivity cost versus new. Prior product required one week for employees to get to productive use versus one day for new product. Old cost = loaded salary (say $2,000 per week) times number of employees (1,000) = $2,000,000 old time to productivity cost. New cost = $400 salary for one day times 1,000 employees = $400,000. Old cost of $2,000,000 minus new cost of $400,000 yields a $1,600,000 cost reduction.
Reduce document costs	Subtract new document distribution costs from former costs. Formerly, customers paid $20 per document for 1,000 documents, for a cost of $20,000. New documents are on-line with the product for a cost of $0. Cost reduction is $20,000.

revenue that it expends. In an organization where technical communication is believed to be a "support" service to the more important development of products and services, it will be difficult for communicators to make the case that they contribute to revenue. When revenues do go up, others will rush to claim the credit, led by the sales force who believe, not surprisingly, that they are responsible for any such increase. It will be difficult, then, for technical communicators to demonstrate how their efforts alone have created the revenue increase. To do this, you have to keep careful records and statistics and link them as well as you can to your group's efforts. Some of the ways that technical communicators have for increasing revenues are straightforward and obvious, especially when the communication group is selling some or all of its products. Table 3.8 provides examples of methods by which technical communicators can increase revenues.

Note, as mentioned earlier, that when revenues go up, many people will claim credit for the rise. In most cases, it will be difficult for technical communicators to prove that we were the cause of the revenue increase. In a case where no other changes have been made to a product and where no special sales incentives have occurred, but where we have made significant improvements to the documents, we might at least get some of the credit. In the case where we develop and sell a product ourselves, such as third-party documents, we can claim nearly all of the credit.

TABLE 3.8	Calculating Value Added Through Cost Avoidance
Cost Avoidance	**Calculation**
Avoid product returns	Measure return rates on similar existing products. For the new product, subtract the lower return rate from the historical rate. For example, 20 percent less 10 percent means we reduce return by 10 percent. If gross sales are $1,000,000 we have avoided $100,000.
Reduce or eliminate installation and administration costs	Measure installation costs for similar existing products. Subtract installation costs, if any, for new product and multiply by the number of units sold. If our products have always required an installer to be sent to each customer's site at an average cost of $1,000, and our new, improved manual means that customers can install themselves, we save $1,000 times 1,000 customers, or $1,000,000.
Reduce or eliminate training time	Measure training costs for similar existing products. Subtract training costs, if any, for new product, and multiply by the number of users who would have to be trained. If our previous product required one week of training at an average cost of $1,000 per employee (including employee non-productive time while in training), and our new product has tutorials built in so no training is required, we save $1,000 times the number of employees who would use the system.
Reduce time to productivity	Compare time required to achieve productive use of the new product versus earlier versions, competitor's products, or similar products. Subtract time required for your new product from time required for compared product, and then multiply by the number of users for a given organization. If it takes users of a competitive product 20 hours to use it productively and our improved documentation system teaches them how to do so in 2 hours, we save 18 hours per person. If a customer has 500 users, we multiply 500 × 18 hours, for a total of 9,000 hours. Assuming a loaded rate of $70 per hour, the avoided cost is $630,000.

Intangible Contributions Technical Communicators Can Make

Here we get into the gray area where we know that the technical communication group is contributing to the organization's bottom line but we cannot directly prove it or calculate it. Sales staff come back from the field and say that the customers are telling them they love the new documents. A comparison review in an industry journal rates the documentation as the best in the field. You get unsolicited calls, letters, or e-mails from customers congratulating you on the improved on-line help system (yes, this does happen). Some of your documents win awards in the local STC competition or, even better, in the international competition.

TABLE 3.9	Calculating Value Added Through Revenue Enhancement
Revenue Enhancement	**Calculation**
Contribute to increased re-sale among existing customers	If you make significant improvements to version 2.0 of a product's document and a much higher percentage of your customer base upgrades to version 3.0 than is normal, you can take some of the credit. If 50 percent of your customers normally upgrade when a new product is introduced, and, after your improvements, 75 percent upgrade, you can claim the difference and multiply by the product price to show how much gross revenue you generated. However, others will also claim responsibility: engineers will say it's the new features they put in, sales will say it's their aggressive upgrade campaign, the advertising agency will say it's their upgrade package, and so on.
Give away some of the documentation and sell the rest	All of the revenue you generate by selling documents should be credited to your group. A dramatic example is the one listed earlier in the cost reduction section. By giving away electronic documents rather than paper-based ones, you can reduce costs dramatically. You can also increase revenues by offering for sale the paper versions of the documents, for those who want to order them. You can create further revenue by creating and selling documents for specialized audiences, such as theory of operations guides and user guides on advanced topics. To calculate revenue, subtract the development costs to generate the documents from the total revenues produced from selling them.
Create vertical market products for niche audience groups	For many products and services, the technical communication group creates documents that are generic and are aimed at general readers or users in any industry. With some types of products and services, however, you can create specialized sets of documents aimed at particular industries. For example, if your product is a point-of-sale system, you might create document sets describing optimum use for department stores, drug stores, grocery stores, etc. To calculate the revenue enhancement, subtract the cost to generate the documents from the revenue they produce.
Create your own third-party documents; compete with yourself	So many computer documents were so bad for so long that third-party publishers (such as Que, the "dummy" series, the "idiot" series, etc.) have made a major industry writing better documents. While you might have to develop fairly standard document sets for shipment with the product, consider developing and selling books similar to what the third-party publishers sell. Microsoft has been doing this for years, delivering a fairly basic set of documents with a product and then selling more detailed versions. To calculate the enhancement, simply subtract the development costs for such documents from the revenues they generate.

While all of these things tell us that we are creating good documents, they are anecdotal and difficult to quantify. Nonetheless, when we are demonstrating how we add value, we need to take such factors into account and be sure to communicate them to the rest of the organization (Redish 1995). It is important to save each of the pieces of information that we receive about such intangibles, including making notes to a file when we get positive verbal feedback.

One of the primary ways in which technical communication groups make a positive contribution is in improving the ethos of a product and an organization. Ethos in this sense means the "reputation," or the "feel" of the product and of the organization behind it, the sense that the product's user gets about the quality of the thing itself and of the organization that created it. With some products, the ethos that they carry is largely determined by what the technical communication group creates. This is especially true for software, where the product is loaded into a computer from a CD-ROM or a file delivered electronically. After that, the customer never sees the software again unless it breaks. Instead, the customer sees the documents and screens in the product's user interface. To the customer, the documents and the screens *are* the product. The programmers who wrote the code have provided an important support service, for which we are grateful. On a Web site, the Web developers, who may be the technical communication group, are completely responsible for the ethos generated by the site and the reaction of customers to it. Unfortunately, many scientific and technical practitioners focus so intently on the practical, functional aspects of what they are doing that they have no sense of the less tangible aspects of the product. They are, in effect, ethos-challenged. In fact, they may see nothing wrong with sending out a beautiful piece of software code on a CD-ROM with no other documentation than a photocopied, typewritten sheet of instructions. How do technical communicators convince such people that our enhancment of the product's and organization's ethos is important? It requires education through constant communication. Publicizing the awards and the anecdotal customer communications with positive feedback will help. The sales and marketing group can also help in project meetings by emphasizing how important good documents are to their efforts.

Technical Communication Public Relations

A technical communication manager should assume that he/she is on a constant, ongoing public relations campaign regarding the intangible contributions that the communication group makes. Every positive communication should be redistributed to the entire product or service team, even the techies, so that they will begin to learn that the technical communication group does affect the reputation of their product. When someone wins a communications award or experiences some other success, you should make as big a deal out of it as possible. Preferably, it should be communicated to the entire organization, through its employee newsletter, intranet, or by whatever means the organization uses to communicate with everyone. Make sure that the communication mode you use will get to higher management (Natchez 2001).

Technical Communication Management Worksheet

Technical Communication Public Relations

Technical Communication Group _____   Month _____

Check each of the following PR possibilities to ensure that you have communicated with other organizations about your group's activities. Add new PR opportunities in the blank lines at the bottom.

	E-mail	Intranet	Memo	News-letter	News Release	Other
Positive feedback from customers						
Positive feedback from sales staff						
Positive customer evaluations						
Positive usability test results						
STC award(s) won by group member						
Other TC awards won by group member						
Group member elected or appointed to STC leadership position						
Group member gives public presentation						
Group member gives paper or presentation at a national conference						
Group member publishes article in a technical communication journal						
Group member publishes other articles or books						
Group member completes advanced degree						

Your public relations campaign is an effort that can simultaneously provide three positive benefits. First, it helps teach people in other groups about what you do and about its value. Second, it constantly reinforces the idea that you are providing value. And third, it acts as positive reinforcement to the entire communication group and especially to those employees who are specifically cited.

The Technical Communication Public Relations Worksheet provides some ideas to enhance the reputation of your group both inside and outside of your larger organization. Checking this list at the beginning or end of each month will help remind you to communicate your successes to the larger organizations to which your group belongs.

References

Barr, J. P. and S. Rosenbaum. 1990. Documentation and training productivity benchmarks. *Technical Communication* 37(4): 399–408.

Bryan, J. G. 1994. Culture and anarchy: What publications managers should know about us and them. *Publications Management: Essays for Professional Communicators.* Edited by O. J. Allen and L. H. Deming, 55–67. Amityvillle, NY: Baywood.

Carliner, S. 1997. Demonstrating the effectiveness and value of technical communication products and services: A four-level process. *Technical Communication* 44(3): 252–265.

Finkler, S. A. 1996. *Finance & Accounting for Nonfinancial Managers.* Paramus, NJ: Prentice Hall.

Frick, B. 2001. Minding your business: Contracts 101. *Intercom* 48(7): 49–50.

Hackos, J. T. 1994. *Managing Your Documentation Projects.* New York: John Wiley & Sons, Inc.

Hart, G. J. S. 2001. Prove your worth! *Intercom* 48(7): 16–19.

Lang, D. and D. Ricks. 1994. "A Documentation Database for Managing Time and Costs," in *STC 41st Annual Conference Proceedings*, Society for Technical Communication.

Mead, J. 1998. Measuring the value added by technical communication: A review of research and practice. *Technical Communication* 45(3): 353–379.

Mirel, B. 1993. "A Study of Instructions for Information Systems: Variations on a Minimalist Theme," in *STC 40th Annual Conference Proceedings*. Dallas: Society for Technical Communication.

Natchez, M. 2001. Managing up: The overlooked element in successful management. *STC Management* 5(4): 2–6.

Polson, J. J. K. 1988. "A Model for Management: Defending Yourself Against Murphy," in *STC 35th Annual Conference Proceedings*. Washington, DC: Society for Technical Communication.

Plung, D. L. 1994. Comprehending and aligning professionals and publications organizations. *Publications Management: Essays for Professional Communicators.* Edited by O. J. Allen and L. H. Deming, 41–54. Amityville, NY: Baywood.

See, E. J. 1995. Moving to an entrepreneurial model: Providing technical information services within a large corporation. *Technical Communication* 42(3): 421–425.

Smith, D. L. 1994. Estimating costs for documentation projects. *Publications Management: Essays for Professional Communicators*. Edited by O. J. Allen and L. H. Deming, 143–151. Amityville, NY: Baywood.

Thomson, T. 1998. Percent-of-total budget forecasting method. Fredrickson Communications. http://www.fredcomm.com/articles/artcl_percent.htm (November 1, 2001).

Zvalo, P. 1999. Pricing a documentation project is part science, part art. Writer's Block. http://www.writersblock.ca/summer99/a-buswor.htm (November 1, 2001).

For Further Information

Steven A. Finkler's *Finance and Accounting for Nonfinancial Managers* (1996, Revised and Expanded Edition) provides excellent explanations of fundamental financial and budgeting concepts for managers in nonfinancial areas such as technical communication. The American Management Association (http://www.amanet.org) offers a good three-day seminar, *Fundamentals of Finance and Accounting for Nonfinancial Managers*.

Edward J. See's article, "Moving to an Entrepreneurial Model: Providing Technical Information Services Within a Large Corporation" (1995), gives ideas and examples for using levels of quality as budgeting and project management tools in the context of larger organizations.

JoAnn Hackos's *Managing Your Documentation Projects* (1994) describes how to develop sound information plans and project plans that can essentially act as contracts for technical communicators. Betsy Frick's article, "Minding Your Business: Contracts 101" (2001), provides basic information about writing contracts for technical communication managers.

Hackos's book also provides detailed information on estimating technical communication projects. Judith J. K. Polson's article, "A Model for Management: Defending Yourself Against Murphy" (1988), discusses the importance of preparing good estimates and provides a formula to help visualize the relationships among quality, functions, resources, and time on a project. Peter Zvalo's on-line article, "Pricing a Documentation Project is Part Science, Part Art" (1999), discusses the differences between fixed cost and hourly projects and estimates. David L. Smith's chapter, "Estimating Costs for Documentation Projects" (1994), describes and provides examples of estimating methods, including the idea that an estimate should divide the deliverables into the smallest segments possible. Hackos, in her book (1994) and Lang and Ricks in their article, "A Documentation Database for Managing Time and Costs" (1994), emphasize the importance of preserving accurate historical information about actual project results to aid in preparing accurate future estimates.

Calculating and communicating the value added by technical communication has been treated extensively. Hackos (1994), Brian, in his chapter, "Culture and Anarchy: What Publications Managers Should Know about Us and Them" (1994), and Plung, in his chapter, "Comprehending and Aligning Professionals and Publications Organizations" (1994), all discuss the importance of technical communication groups aligning themselves with the goals of the larger organization as a way to help show value added. Geoffrey J. S. Hart, in his article, "Prove Your Worth!" (2001), distinguishes between the quantitative and qualitative values that technical communicators add. Jay Mead, in "Measuring the Value Added by Technical Communication: A Review of Research and Practice" (1998), provides a survey of research and some basic considerations for conceptualizing value added. Saul Carliner, in "Demonstrating the Effectiveness and Value of Technical Communication Products and Services: A Four-Level Process" (1997), describes a comprehensive process for visualizing, calculating, and reporting value added. Meryl Natchez, in her article, "Managing Up: The Overlooked Element in Successful Management" (2001), stresses the importance of using public relations techniques to improve the perception of technical communication value added and of using the appropriate communication modes to ensure that the message gets to upper management.

Questions for Discussion

1. Find a balance sheet for a Fortune 500 company or for a company in your geographical area. Annual reports for publicly held companies contain a balance sheet, although it may be labeled something else. Do you see any ways in which a technical communication group might be considered an asset? Where does the technical communication group fit on the liabilities and capital side of the balance sheet?

2. Obtain an income statement for a Fortune 500 company or for a company in your geographical area. Does it have a separate line for technical communication? (Some do.) Where do you think the TC group is accounted for on the statement?

3. If you work in a technical communication group, see if you can get a copy the latest monthly budget for the organization to which technical communication reports. If you do not work for a technical communication group, see if you can get a budget by asking someone who does, or by contacting members of the nearest STC chapter. Which lines on the budget do technical communicators contribute to? Revenue? Expenses?

4. Find an example of a service-level agreement that you subscribe to. Good places to look include your local telephone service, long distance service, cellular service, cable/satellite TV service, automobile maintenance program, and others. Why did you choose the service level you have? For customers, what are the advantages of service-level agreements? The disadvantages?

5. Consider the 14 variables in this chapter that can affect the cost of a technical communication job. Which ones are most likely to affect a project where you work (or where you would like to work)? Are there other variables that are not presented here?

6. Would you rather do a technical communication job on a fixed price or an hourly charge basis? Why?

7. Should a manager automatically pad all estimates or report on exactly the amount he/she needs?

8. Consider the formula and examples in this chapter for measuring value added by technical communicators. Can you think of other ways that technical communicators can reduce or avoid costs or add to revenue?

9. Why should a technical communication manager have to embark on an ongoing public relations campaign to show the value of the TC group when other groups do not have to do so?

CASE 3

Value-Added Calculation

The Management Situation

While your manager at Aardvark Enterprises has been very supportive of your fledgling technical communication group, word has come down from higher management that every group in the entire organization must prepare a "value-added statement" that shows how they contribute to the bottom line and how they plan to do so for the next year.

Your boss is somewhat concerned that technical communication might merely be seen as an expense, but, not being a member of our discipline, does not have many ideas about how to show value added for a technical communication function. You, then, must prepare a report demonstrating how the technical communication group adds value. Because your group is new, you cannot show any past value-added measures, but you can show how you plan to help in the coming year.

While upper management has not said that they plan to outsource any of the company's internal functions, you believe that you should prepare your report as a defense against that possibility.

Some background information:

Aardvark is a small company with about 200 employees, 120 of whom are engineers and programmers. Aardvark designs and develops both hardware and software products to support point-of-sale systems, a highly competitive niche market. A secondary product line is financial analysis software. Aardvark's products have consistently been better designed than competitors' products, but customers have reported frustrations with learning how to use them, and returns have been high.

Up until now, all documents have been prepared by people who were not technical communicators. The documents are product-oriented and are largely descriptive of the product's features and functions.

The company currently sends only paper documents, with an average cost of $7.00 each for some 50,000 documents per year.

The company hotline receives 20,000 calls per year with an average cost per call of $28.00.

The company's large financial software program requires that Aardvark engineers install it at customer sites. The engineers believe that customers could do it themselves if someone could write a good enough 50-page installation manual, but the engineer's attempts to do so have not been successful. Each installation trip costs $4,000. Projected sales for the product for the next year call for 300 units.

Aardvark's last attempt at a personal investing program led to a 40 percent rate of product returns, with many customers stating that they could not figure out how to use the software. The new personal investing program, Midas, is slated to sell for $100 and has projected sales for the next year of 100,000 units.

The sales/marketing department believes that it could sell sets of vertical-market documents to owners of small, local banks and brokerage houses. The document set would have to be about 100 pages long and would sell for $100. They believe that they could sell 5,000 units.

You were hired to start a technical communication group at Aardvark largely because of customer complaints about the quality of previous documentation. There is a perception among upper management that Aardvark has some "image" problems because it has sent out cheap, unprofessional documents in the past, leading to high rates of hotline calls and customer returns.

The Assignment

Write a report showing how the technical communication group will add value in the coming year. While it is important to show that you will add as much value as you cost, you do not want to overstate the case or to use dubious estimates. With a new technical communication group, you are going to have some acculturation and learning requirements that will lower the group's productivity in its first year. You are also aware that you may be held accountable for any estimates about cost reductions or revenue enhancements that you make in the report.

The report is going to upper management. You therefore want to show much of the information in tables or charts rather than in text. You also realize that this is an excellent opportunity to educate your own supervisor and upper management about the various ways in which the technical communication group can contribute to the company's mission and goals.

Helpful Hints

Use the formula presented in this chapter to calculate the value added.

Resist the temptation to make exorbitant claims about how much you can reduce costs or increase revenues.

For the articles presented in a special issue of STC's *Technical Communication* on value added, see http://www.techcomm-online.org/shared/special_col/measuring/menu.html

For Carliner's (1997) article on value added, see http://www.fredricksoncommunications.com/articles/value/abstract.htm

Evaluation Criteria for Case 3

Does your report take into account the rhetorical situation? Does it have the proper structure and tone for a management audience?

Does the report clearly show the calculations behind each projected contribution to reducing costs or enhancing revenues?

Are the projections reasonable for a new technical communication group that will be in the process of being formed in the same year for which the projections are made?

Does the report address ways that technical communication adds value other than positive effects on costs and revenues?

CHAPTER 4

Management Communication

Introduction

Communication skills determine how a manager is perceived by direct reports, peers, and supervisors. Those skills will also, therefore, determine how the manager's function and employees are perceived. A highly regarded manager elevates the reputation of his/her group and function, while a poorly regarded manager causes others to think poorly of his/her group also. How the manager and the group are perceived is largely determined by how they communicate with others. Further, a manager and a group define their principles for others by how they communicate.

A manager who claims to engage in principle-based leadership must be very careful about how he/she communicates. In one of my annual 360-degree performance reviews, I was rated low on "trust" by my peers. This stunned me, as I took pride in telling the truth, even when it was painful to people. What I learned after investigating further was that I always said yes to any request that a peer made of me. When I was subsequently not able to fulfill all of the requests, the other managers no longer trusted that I would do what I had said I would do. It was not a matter of intention at all, but rather a matter of results. They did not believe that I was intentionally dishonest but, nonetheless, they could not trust what I said about deliverables. After that, I said no much more often. My popularity did not go up any, but my ratings for truthfulness did. Hence, the perception by others regarding one's principles can differ from one's own intentions, even when the intentions are positive. Technical communication managers should consider all communication situations as potentially communicating their values and principles to their employees, peers, and managers.

Management communication is a major area of study, with its own set of literature and numerous books dedicated to it. This chapter will not investigate that body of information, but will focus instead on proven methods for technical communication managers to use to communicate more effectively with their groups, peers, supervisors, and others.

Many technical communication managers naïvely assume that they do not need to think about communication skills—that, *a priori*, because they are communicators they have good skills. While this may be true as it applies to user manuals or to government reports, the skills that are used to develop such documents are necessary but not sufficient to ensure adequate communication in a management role. So, technical communication managers should attend carefully to how and why they use various communication methods.

Rhetorical Situations

Have you ever wondered why you go home at night so tired, when you haven't done much "work" all day. In fact, often it is those days when you seem to get the least done that are the most fatiguing. One of the major factors contributing to that fatigue is the necessity for communicating with multiple people on multiple occasions during the day. Each communication with each person or group presents another rhetorical situation. That is, you must decide for each communication what your message is, who your audience is, what your purpose is, and which communication methods will most effectively communicate that message for that purpose to that audience. Much of the time you make those rhetorical decisions instinctively and automatically, without thinking (which, obviously, can lead to problems). Nonetheless, you are expending energy each and every time you communicate with someone. Technical communicators, who tend to be more aware of customizing their communication modes for each situation, are likely to expend more energy at it than most. Further, technical communication managers are likely to have to change rhetorical situations dozens and even hundreds of times in a single day. Figure 4.1 shows the complexity of the communication situations with which a second-level manager might be confronted daily. If you are a first-level manager, you can simply cover the bottom row.

Notice that the technical communication manager has three main groups with whom to communicate: (1) employees in his/her group and elsewhere, (2) peers in his/her organization and elsewhere, and (3) upper management, again in his/her organization and elsewhere. Each of those groups involves different rhetorical considerations when communicating with them. Further, each individual in each group may require different communication modes.

Communication theorists often talk about how "rich" a communication medium is. By rich they mean how well the medium/method leads to participation, generation of ideas, speed of decision-making, etc. It should be a technical

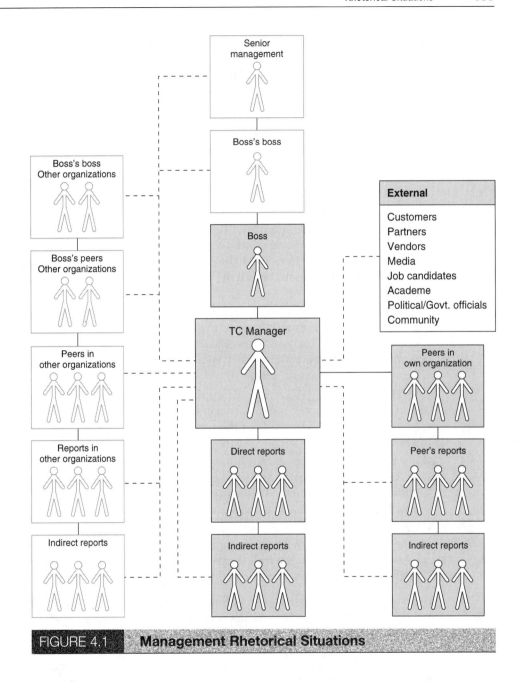

FIGURE 4.1   **Management Rhetorical Situations**

communication manager's goal to make each communication as rich as possible. To do so requires that we consider a number of possible variables and that we try to optimize each of those variables for the given communication purpose and audience.

Communication Modes

You can use one of several communication modes to deliver any piece of information. You have to decide for each type of communication or each purpose which of these modes to use.

One to One—This is the most common mode, one that most of us use dozens of times every day. We usually make transitions from one-to-one communication situations easily. However, with some people, such as problem employees or with the president of the organization, it may not be so easy. In general, one-to-one communication should be used when the message you have is for only one particular person, or when you want to modify some more common piece of communication for that person individually. One problem with one-to-one communication is its variability. That is, you may think you have just delivered the same message to each of ten people individually, but if you survey them you may find that they heard strikingly different messages.

One to Many—All managers will occasionally use the one-to-many mode, usually to transmit information about organizational matters to everyone in their group, or about a certain project to everyone who is working on it. One-to-many communications are effective when you want to ensure that everyone gets the same message, although whether they interpret it in the same way is another story.

One to None—We have all seen managers who unintentionally use this mode, perhaps just to hear the sounds of their own voices. Of course, you don't want to use this one.

Many to One—This mode is fairly rare. However, it does have its purposes. Sometimes it is a good idea to ask your group to talk about a subject while you listen without commenting. This is especially true if you want them to talk about problems that they might not otherwise be willing to discuss.

Many to Many—A group discussion, even though only one person speaks at a time, is a form of many-to-many communication. Another increasingly common form is teleconferences with groups in two or more meeting places communicating via the Internet or dedicated lines, using audio and/or video.

The overwhelming majority of a technical communication manager's discussions will occur using the first two modes, one to one and one to many. E-mail enables us to use a one-to-many, remote communication mode. Ask yourself, though, if the type of information you are communicating would be better delivered in person in a group meeting, or by discussing it with each person individually, no matter how inefficient that might be. Also, keep in mind that any e-mail message you send can end up in the public domain.

With each communication situation you encounter, you must decide which mode will work most effectively.

Communication Locations

Communication begins before it begins. Part of the message you send someone when you communicate is associated with all of the environmental conditions surrounding the interaction before the first word is exchanged. If you call someone who is 50 feet down the hall on the telephone and ask him or her to come to your office, you have already affected the message that you are going to deliver, before you begin delivering it. If you walk down the hall to the other person's desk to deliver the message, you have begun the communication in a different manner. It is worth considering where you conduct communication with others, particularly with those who report to you. You have several obvious choices:

Your Office—Most of us are most comfortable in our own spaces, whether they be at work or at home. For a manager, the office is often a symbol of power and position. It is a place where the manager feels "safer" communicating with people, especially if the communication task is difficult or complex. However, think about this from your employees' point of view. For many employees, the manager's office is a place to fear and to avoid. Bad things happen there. People get reprimanded, grilled, fired, etc. Even if you don't do those things very often, your office is still not a relaxing place for many of your employees to be. You are much more likely to conduct true two-way communication somewhere else. I had one boss who gave performance reviews in a neutral place—a meeting room or the cafeteria during off hours—which resulted in a more relaxed, open atmosphere, without quite as much of the aura of hierarchy and power associated with sitting in his office. Obviously, reprimands and other negative feedback should be conducted privately, but otherwise it is a good idea to communicate with people in their own workspaces or in neutral ones.

Their Office—Most people are indeed more relaxed in their own workspaces, so that real communication works better there. This is especially true when you are simply trying to get a progress report or to listen to problems that you might help fix. If you believe that a communication session is about to turn negative or that you are going to have to correct someone's performance, you should move the session somewhere else that is more private. Giving positive feedback in front of others is sometimes part of the reward. However, some people would prefer to get such feedback privately also, rather than in front of others. For some people, receiving praise is very stressful (believe it or not), so it is a good idea to consider whether you should give praise publicly.

Meeting—Meetings are a logical place for group communication, but not a good place for negative reinforcement, either with individuals or with the entire group. Castigating or grilling one person in front of his or her peers is a poor way to manage. While it may engender fear, it will not create respect. Keep in mind that any information you communicate in a group meeting will be considered "public domain," even if you tell people not to talk about it. Also keep in mind that people in the group will interpret what you say in their own way, informing it with their own preferences, biases, fears, etc. If you are presenting an uncomfortable communication to a group, it is often a good idea to repeat it and/or to give out a written handout to reduce the number of individual interpretations. Some kinds of positive feedback are appropriate for a group meeting, especially recognizing innovation or creative problem solving. However, some kinds are best done in private, especially when you are praising someone for improving their performance from an unacceptable level to a higher one.

Hallway—In the time pressure of the normal work day, it is not unusual to meet someone in the hallway or by the water fountain and to remember that we needed to talk with them about something. If the matter is relatively trivial, the chance environment is probably fine. However, if it is something more substantial, it is better not to try to communicate on the fly, but rather to use the opportunity to set up a time later in the day when you can continue the communication. Many people do not respond well to unexpected communication, especially if it is of a complex or controversial nature. They may feel ambushed and/or threatened. Again, for real communication to occur, chance meetings are usually not a good time.

External—We can also communicate with people in off-site locations. One of the most effective ways to get a group to brainstorm and to think conceptually is to get them away from the workplace. I have been in group meetings in hotel meeting rooms, racetracks, picnic grounds, homes, bowling alleys, restaurants, and, yes, bars. Being away from the day-to-day workplace, the telephone calls, e-mail, kibitzing colleagues, and other distractions helps employees relax. The environment itself helps get the idea across that this is not work as usual and that different modes of thinking and interacting are appropriate.

The way space is arranged in an organization affects how people set up communication networks and how they interact with others (Beck 1999). Spatial arrangements convey the status of employees and give advantages to some kinds of communication over others. One of the major problems with the traditional, hierarchical organizational structure is the distortions it causes to information as it gets passed up and down the chain of command. Even organizations that are trying to eliminate such hierarchies sometimes unwittingly encourage them with the spatial arrangements they set up. With the increased ability of all employees to access the same information, many organizations are flattening the number of management levels they have and are also experimenting with more fluid, flexible spatial arrangements. Such arrangements allow for "instant" communications networks to be set up, especially for project teams whose members change frequently. A technical communication manager should consider the best ways to arrange co-workers so as to enhance their communication abilities.

While it is something that most of us rarely think about, where we communicate is often a big part of communicating our message. By being in the wrong place, we can alter how people respond to what we're saying, and by being in the right place, we can enhance it considerably. It is worth spending some time before communicating difficult, complex, or uncomfortable messages to consider what the right environment is.

Communication Times

As with the place where we communicate, the time when we do it often becomes part of the message. The atmosphere at an eight o'clock Monday morning meeting will differ from that at a four o'clock Friday afternoon meeting. When it comes to timing, we have three main factors to consider.

The Time of Day—According to pop psychologists and to more empirical measures such as brain wave monitoring, there are "morning" people and "afternoon" people. If our boss is one of those people who needs three cups of coffee and an hour or two of solo work to get going in the morning, it is obviously a bad idea to charge in at 8:01 with a radical proposal. If we have an employee who charges through the morning getting lots done, but who saves the routine tasks for late in the afternoon, 4:30 is the wrong time to expect him/her to help us figure out how to make some innovative changes.

The Day of the Week—Some of us seem to have weekly as well as daily biorhythms, moving from manic Mondays to more frivolous Fridays. Other people seem to work more steadily, regardless of the day. Still, it is worth paying attention to the mindset of people with whom we want to communicate as it relates to the day of the week. If our boss likes to wrap up loose ends on Friday afternoon and to go home with something of a "clean" slate, it is not a good idea to rush in with the latest, unsolvable crisis at 5:30.

During or After Work Hours—While we would usually prefer to conduct all communications during regular working hours, it can sometimes be advantageous to conduct them at other times. Sometimes people have brainstorms over the weekend that they simply can't wait until Monday to present. While we should all strive to preserve our private lives, spending a few minutes working with a colleague over the telephone or via an e-mail exchange in the evening or over the weekend can be rewarding. I once had an employee who was shy and introverted and who had a difficult time discussing personnel-related issues at the office. We had several very good telephone conservations in the evenings where he felt freer to open up and to discuss difficult issues and to work on them. Another favorite after-hours communication time is the Friday afternoon happy hour, although that is probably a time for managers to be wary of discussing anything too weighty.

In any case, the timing of communication often combines with the environment to affect the message that is being delivered. Managers should take this into account, especially when they are communicating information that is unusual, difficult, or challenging.

Communication Media

Of course, the medium by which we communicate greatly affects the content of the message. Whether you agree with McLuhan (1965) that the medium is the message, it is undeniable that the medium has a major impact on how the message will be received and what content will be drawn from it. A manager has numerous choices for media, and choosing the correct one can be critical to communicating successfully (Trevino, Daft, and Lengel 1990).

Direct Conversation—This is by far the most common medium, although one that is perhaps waning with the increase of e-mail and other electronic media. People have certain minimum expectations about being able to discuss things in person with their supervisors. I knew of one supervisor who informed his group that he was too busy to meet with any of them personally and that all future communication would occur via e-mail. He of course became the object of ridicule and scorn. Even with the ease of e-mail and other media, managers should consider which communication tasks are better done in person. It is much easier to say no to an electronic proposal for a radical change than it is to say it to someone sitting right in front of us. And, while it may be tempting to give an employee negative feedback through the veil of e-mail, it is much more appropriate and effective to do so in person.

Memo—The hard copy, paper office memorandum is being replaced almost totally in many organizations by e-mail. Many of us will be thankful for this, but paper memoranda have their place. First, positive reinforcement, such as the announcement of an award of some kind, seems more formal and more "permanent" if it is done via paper. Contracts or service agreements on paper, with signatures, give a project more legal and psychological authenticity than an electronic version. And paper memoranda about personnel-related issues can help to communicate a stronger sense of concern or urgency. All e-mail messages should be considered public because many of them become public whether their sender intended it or not. Recent court rulings have allowed e-mail backups to be subpoenaed, both in civil and criminal cases. For many communication purposes, a single memorandum sent to one person is a much more secure way to communicate.

Letter—Even more formal than a memorandum is a paper letter. Communicating with employees via letter seems almost absurd, unless it is about a serious personnel matter. Nonetheless, there may also be times when a letter is more appropriate than any other medium. For reward purposes, a letter arriving at the employee's home, especially if it includes a check, can be a very powerful statement. A formal letter spelling out a personal action plan (see Chapter 2) puts an employee more on notice than does an e-mail version of the same message. And, of course, contracts and proposals with external organizations are usually accompanied by a letter.

Phone (including voice mail and beepers)—The ubiquity of the telephone in many people's lives may make them easy to communicate with, 24 hours a day, 7 days a week. However, a manager should think twice about taking advantage of this medium. We have all seen people roll their eyes as they receive a phone call from their boss while they are trying to have a relaxing Saturday evening dinner. The boss should be having a relaxing Saturday evening dinner too, rather than bugging employees. But even if the boss cannot relax, he/she should consider leaving employees alone during the times that they spend with their families and friends away from work.

An important aspect of telephone calls is that they are often intrusions into something else that the recipient would prefer to be doing. Further, the recipient may not be ready to discuss whatever subject you wish to introduce. Therefore, telephone calls about difficult or complex issues should be scheduled, so that all participants involved have the information they need and have arranged for the proper amount of time for the discussion. It is not a reasonable expectation that the person you are calling will drop everything else they are working on to talk with you for as long as you want. Other media, such as e-mail, can serve as a preliminary introduction to the topic, with a request in the e-mail for a time when you can have a lengthier discussion via the phone.

This may sound like heresy coming from someone who worked in the telecommunications industry for many years, but business telephone calls are often an inefficient, ineffective means of communication, especially when they are unscheduled. Because a large percentage of business calls are answered by voice mail systems, we end up with labyrinthine systems of messages, responses, responses to responses, etc. Direct conversation and e-mail are often better choices. When you call someone who is 50 feet away from your office and talk to them using your new speakerphone (with its accompanying echo), you send them more of a message than you intend to.

E-mail—E-mail has become the most common means of communication in many fields, and for good reason. It is an especially felicitous medium for technical communicators to use, as many of us are more confident and adept at using the written language than we are in meetings with individuals or groups. Using e-mail wisely can greatly reduce the time required for communication because it is considerably faster and more efficient than paper communication or telephone calls.

However, it is important for managers to be aware of some of the inherent problems with e-mail. First, exchanging e-mails with employees is indeed fast, but it can also lead to misunderstandings, especially because e-mail messages often sound more curt and blunt than their authors intended. An innocuous message to one of your employees can sound too blunt and even mildly insulting, even when you had no intention of conveying anything negative. While it is not common practice to do so, I prefer to put the person's name on a separate line at the top of each message, and to conclude with a pleasant closing, like "thanks" or "cheers." Such minor changes help prevent the e-mail from seeming too curt. Another possible problem with using e-mail regarding sensitive subjects is that it can become public too easily. It is not a good idea to send anything in an e-mail that you would not want everyone else in the organization to see. Computer administrators have access to all of it. An angry employee can forward it wherever he/she wants to. And you can inadvertently reply to someone on a mailing list and send the message to the entire list rather than the lone recipient you intended. While etiquette rules are still being developed for e-mail, a good rule of thumb is not to say anything in an e-mail message that you would not say to someone's face.

Many e-mailers seem to believe that all of the rules for good written communication go away when one is using e-mail. Technical communicators know the importance of the use of clear, concise language and of the way that it is laid out and presented. We should put as much effort into making our e-mail messages clear and effective as we do other types of communication. That extends to putting in paragraph separations and even headings. With the increasing use of html text for e-mails and the ability to attach formatted documents to a message, there should not be an excuse for sending someone a random set of facts and observations, without any order or structure. Further, whether we like it or not, other people in our organizations judge us by how well we communicate with them. They assume that if our communication with them is shoddy and full of organizational and grammatical errors, that our documentation must also have some of the same problems.

Internet—We can also communicate with our groups and with others in our organizations by way of the Web. A group Web site can provide many advantages. First, as the manager of the group, you have a place where you can post important information that everyone needs to see. Second, the site can serve as a training ground and an experiment "bed" for you and your employees to test new techniques and tools. Third, the site can serve as a kind of "advertisement," both with internal and external groups. It can list the types of services you provide and can also list the products you have created. It can also serve as a place where group members and others can find templates, guidelines, and forms that help them do their jobs. A Web site can also improve group morale by providing a sense of public identity and professionalism. Links to the STC, IEEE-PCS, SIGDOC, and other professional organizations demonstrate to others that we are part of an active, dynamic profession (and in many organizations, other employees need to know that). Such links also help our employees search for information.

Another communication resource on the Web involves group mailing lists and groupware applications. These can provide a group or a project team with excellent opportunities for collaborating on documents, exchanging files, maintaining schedules, and more. Some of the groupware sites have very powerful project management capabilities. Because the companies and products in the field are changing so rapidly, we will not include a list of the current ones. Search on "groupware" or "project management" and you will find many such sites and programs.

Finally, instant messaging provides a way for employees to communicate with one another and with external personnel. Instant messaging, available from America Online, Microsoft, Yahoo, and others, provides a window on the computer screen that allows the employee to exchange messages instantly, provided that there is an open line to the Internet on the computer. This can provide significant benefits for a project team, who can ask questions and get answers in "real time" instead of waiting until next week's meeting or sending an e-mail and waiting for an answer. It also provides a boon to groups who are working on projects from multiple geographical locations. While conference calls might be expensive and awkward, instant messaging can provide constant communication among all of the team members.

Video/Audio—Video and audio teleconferencing provide the ability to communicate with remote employees. We can include in a group or project meeting someone who works at another location or who telecommutes. There are many video teleconferencing options, ranging from buying a $50 video camera for a computer to packaged systems running on very fast data lines and costing tens of thousands of dollars. With increasing broadband speeds available for Internet access, we can also use the net for teleconferenced meetings. This can be especially helpful when technical communicators need to share graphics, drawings, photographs, etc.

In some organizations, upper management uses video to communicate with large groups of employees simultaneously. While technical communication managers would rarely use video in this way, we should consider it as an aid to communication among project group members who are increasingly likely to be working at different locations and telecommuting.

Communication Contents

Each profession has its own unique communication needs, and technical communication is certainly no different. It is a good idea for a technical communication manager to assess the basic content of a message before attempting to communicate it to people. While many of the communications tasks of a communication manager will be similar to those of any manager, some of them are peculiar to our field.

Informal—We all engage in informal communication every day. Managers, no matter how busy, should make it a point to do so, especially with the people who report to them. A few friendly words from one's boss can make a big difference in how one perceives the work environment. And, it is especially important for technical communicators and their managers to work hard at communicating informally with the scientists, engineers, and programmers with whom they work. It is too easy for a technical communication group to let itself become isolated because its members do not share the educational and professional backgrounds of their colleagues. That is all the more reason that they should make it a point to communicate and socialize with the technologists, including communication about things outside the work place (DeGraw 1993).

Information—Of course, all communication involves information. Technical communication managers regularly communicate information to their employees related to numerous subjects, including projects and their status, processes and procedures that will affect workflows, the results of document reviews, and performance-related matters. Each of these kinds of content has its own inherent communication requirements, and a communication manager should work hard to ensure that the time, place, mode, and methods chosen to convey information are appropriate.

Top-down—As organizations flatten management levels and as communication systems become more efficient, top-down information being passed from one management level to another, down the chain of command, is becoming less frequent. Nonetheless, it is sometimes necessary for a technical communication manager to report "top-down" information to employees. What do you do if you find yourself disagreeing with the new policy you are being required to present to your employees? Upper management expects you to report it objectively; in some cases they expect you to do so enthusiastically. Your employees expect you to be honest with them. This is a difficult dilemma requiring one to confront principles and ethics. My own policy when this happened was to report the information, as required, but to make it clear that I neither supported nor condoned it. If upper management found out, I would be in trouble. However, it is better to maintain integrity with one's direct reports than it is to appear to be a lackey of upper management. If I disagreed with a policy and this dilemma arose, I usually reported to my boss why I disagreed and what I thought the negative consequences of the policy would be, in the hope that the information would go back up the line. More often, communication from top down is more innocuous and can be passed along fairly easily.

Project Status—One of the most common types of communication for technical communication managers is the status report. Such reports might be made to individual project teams about the status on that single project or to technical communication and other management regarding multiple projects. It is important for technical communicators to report status using the same methods that others do. If the technical groups report their status using Microsoft Project-generated Gantt charts, then the technical communication manager should do the same. Preferably, the communication manager should get technical communication milestones and due dates included on the overall project Gantt chart. If the technologists use PowerPoint slides to report status, then so should the technical communication manager. In fact, a good goal for communication managers in reporting status is to assume that their communications should be as good or better than anyone else's. We are, after all, the communication experts and we should be able to communicate clearly and unambiguously regarding project status.

Performance—Communication about performance should go on throughout the year, rather than being reserved for a single performance meeting at the end of the year. For more, see Chapter 2, under Performance Review Meeting Outline.

Assignment—How a manager communicates a job assignment can have a significant impact on how the employees perceive the work they are being asked to do. The manager can make the job sound routine and boring or exciting and challenging. Even when assigning maintenance work, a technical communication manager should look for a way to make the assignment interesting and challenging. Even with a limited budget, we can usually find something to improve in a document that is being updated. We can find some way to make fundamental improvements that will interest the person working on them. Whether the assignment is a legacy or a new document, we should provide the communicator with as much detail as possible about what kind of outcome we want. When we give assignments, we communicate the true vision and goals of the organization, no matter what the official vision and goals say. We should be able to make the assignment fit those overall goals as well as possible, so that employees can see how their efforts will contribute to achieving them. Consequently, it helps to explain to the employees, especially if they are relatively new, where the product or service they are working on fits within the organization and how its success will contribute to organizational goals.

Vision—A technical communication manager should communicate vision when giving assignments, but it is important to do so at other times, too. We should make sure that all of our employees know how we participate in and contribute to our organization's goals. We can do this during mentoring and coaching sessions, group meetings, assignments, objective-setting meetings, and performance reviews. It is important that we put our employees' work into the larger context of the overall organization.

Knowledge Management—Increasingly, technical communicators are going to be involved in knowledge management. Some have argued that our roles will move from creating content to serving as gate designers and keepers for databases of information that can be accessed for many reasons, such as answering procedural questions, troubleshooting questions, capabilities questions, warranty questions, and others. (Weiss 2002). Already, many companies are providing customer documentation and support on Web sites using such information management methods, rather than publishing books, CDs, or other material. Additionally, many organizations are devising knowledge management systems internally, in an effort to preserve and use the "best practice" capabilities and solutions that employees have discovered and used on previous projects. Such valuable information is often confined to one geographical or administrative area or is lost altogether. Technical communication managers should learn about how knowledge and information management systems work because we will want to be among those who help to design the "data" and the various gateways to accessing and using the data.

Communication Reasons

The reasons we communicate are often complex and interlinked with many other variables. However, a technical communication manager should consider why it is necessary to communicate about a given subject. Much of the time the "why" is automatic. You need to find out about project status or you need to pass along information from human resources. However, it you are having trouble formulating how to communicate a certain message, you should first consider why you want to do so. Some of the more common reasons for technical communication managers include the following:

Befriend/Ingratiate—This is the most common form of communication, usually based around hallway greetings and small talk. You have to assume that most of this is genuine, although you realize that, at least with some employees, it is overdone and moves from collegiality to toadying.

Motivate— Technical communication managers often need to motivate employees, especially those who are working on extremely difficult assignments or, conversely, those who are working on extremely boring assignments. Technical communicators are often motivated by an image of the finished product, what their document will look like when it is completed. They can also be motivated by the idea of solving complicated problems associated with conveying the appropriate information. They are not as motivated as others by the promise of money. One of the STC salary surveys showed that technical communicators were more motivated by public recognition than by salary increases.

Reward—Another reason for technical communication managers to communicate with employees is to provide positive feedback, whether it be a minute or two of praise about a draft or a public award at a group or project meeting. This is especially important, given the fact that writers often go for months without finishing a document. (See Section 3).

Punish—Similarly, communication managers need to give negative feedback when it is required. It is far more effective to give such feedback frequently in small doses rather than once a year in a performance review. The closer the negative feedback is to the event that warranted it, the more effective it will be. (See Section 3 for more.)

Inform—In hierarchical, top-down structures, information passes down from upper management through middle management and then down to first-level supervisors, who act as information conduits. In such organizations, a big part of a manager's job is informing employees about management decisions and developments. While such structures are being replaced with flatter organizations, managers must still occasionally communicate to inform.

Control—Like it or not, sometimes managers communicate to control. This is something to avoid, but a technical communication manager may have to rein in a writer who is insisting on doing things differently from everyone else on a project team, or who is working too slowly. Such communication is more effective if it concerns specific, detailed examples that show employees the impact of their actions rather than a set of general statements, which might not get through or might simply sound like a personality difference.

Build Teams—On a more positive note, technical communication managers often communicate to help build project teams. Especially in the early stages of team building, frequent management communication can help the team establish its goals and begin to put in place its processes. Increased communication helps teams to be successful, whether they are teams of technical communicators, project teams, and/or cross-functional teams. To make such teams successful, it is important that a manager communicate frequently with the team and that the manager provides as many communication resources to the team as possible, including technologies such as mailing lists, instant messaging, e-mail, etc.

Brainstorm—Another frequent communication mode is brainstorming. This is especially important at the beginning of a project when the technical communication team is trying to figure out what types of documents should constitute the "library" for a product or process. While brainstorming can be accomplished in meetings called for that purpose, some of the best brainstorming sessions are the ones that occur when ad hoc groups congregate in someone' s office and start throwing ideas around. See the Meetings subsection for more on brainstorming.

Innovate—Technical communication managers often need to help employees innovate, to work together with them to try to find new and improved solutions for effectively communicating information.

Communicating with Employees

There are numerous rhetorical complexities inherent in communicating with employees. Managers must think much more about how they communicate in a given situation with a given person than they had to before they were in management. Perhaps you were known before you became a manager as a "straight shooter" or as someone who is always upbeat and optimistic or as someone who is wittily dry and sardonic. For the most part, you could use your preferred communication method consistently. When you become a manager, though, that consistency simply will not work. Different employees have widely varying communication styles and preferences, and different situations require using modes with which you may not always be comfortable. Giving someone a rough performance review, for example, while cracking jokes and trying to keep things light, will undercut the important communication you are trying to achieve.

It is important for a manager to consider the preferred communication methods for each direct report (McGuire 1991). While you cannot possibly know the communication preferences of everyone in your entire organization (unless it is very small), it is necessary to know those preferences for your own employees. You have to communicate with them about a range of issues, many of them serious and sensitive. Trying to do so without considering how to get the message across most effectively will cause many problems. Some people find some types of communication uncomfortable and even intimidating. You are not likely to communicate with that person successfully using one of those methods. While there may be times when you must use a particular method no matter how uncomfortable it makes someone, in general it is far wiser to try to use the communication modes with which they are most comfortable.

How do you assess what your employees' communication preferences are? There are two main methods. First, ask them. While this may seem patently obvious, think about how many times in your career you have been asked by a manager how you prefer to communicate. When you ask employees this question, some may be so stunned that they don't know how to answer. You may have to prompt them with more specific questions, such as "Do you prefer communication in person, via e-mail, via the telephone? Do you prefer inductive or deductive approaches—do you want the big picture first or the details first? Do you prefer brevity or breadth? Do you prefer some small talk or chitchat, or do you want to get right to the subject? Do you prefer frequent, short communication, or less frequent, longer sessions? How do you prefer to get good news? Bad news? Do you prefer to communicate early in the day or later in the day?" Such questions should help you understand how and when you can best communicate with each employee.

The second method is to use one of the personality surveys, such as Myers-Briggs, to determine employee's (and your own) communication preferences. While many people are far too willing to believe in the absolute validity of such tests and too cheerfully label themselves IBDAs (or whatever), such surveys can help us find out about preferences. Searching on the Internet on "Myers Briggs" or "personality type" will yield numerous sites with additional information, including sites where you can take the survey on-line. As mentioned in Chapter 2, having your group take one of these surveys together can be enlightening and can help lead to better group unity, as everyone becomes aware of the communication preferences of everyone else (Hackos 1990). Another valuable testing tool is available at http://www.strengthsfinder.com. This test provides data about each employee's five main strengths, which can help you figure out the best methods for communicating with and rewarding them.

Once you have determined people's preferences, you can try to tailor the most important communications with them to fit. Obviously, some types of messages will need to be transmitted by less taxing, more efficient methods. Information that is especially important to each individual, such as performance and financial information, however, should be delivered in a manner that will ensure that the messages get across.

You must also consider the medium you use when communicating with employees. As discussed in the section in Chapter 2 on problem employees, there is a hierarchy of formality associated with different communication media. Most of us would prefer, for example, not to receive a sealed envelope from our boss on Friday afternoon.

Communicating with Peers

Communicating with management peers is significantly different from communicating with direct reports. Communication among managers is based much more often on solving problems and facilitating progress toward goals than on dealing with the more detailed concerns that employees who are developing documents or products must be concerned with. There are two kinds of peer managers: those within one's department or division and those in other areas. It is important to work well with management peers within a department. Many technical communication managers need to work much harder at developing good relationships with management peers in other areas, such as marketing, engineering, and customer service. Those peers can be critical allies in helping technical communication get adequate resources and helping get us assigned to project teams early enough to do quality work.

For the purpose of communicating with peers in other departments, it is important to notice the communication modes that they use. If everyone else is on a mailing list or an intranet project site, you need to join and participate also, even if you don't say too much at first. If everyone else brings along an overhead or two to each project meeting, you need to do the same. You need to use the same methods that peers use, even if you might see other communication methods that would be more effective. Start by using the common, accepted communication methods and then try the ones that you think might work better. One unfortunate reality of communication among primarily scientific and technical employees is over-reliance on meetings. Because such employees are often not good at and not confident in communicating with the written word, they choose to do so in meetings, which can seem wasteful and interminable to a technical communication manager, who believes that much of the communication could have occurred in a few e-mail messages. Nonetheless, you need to use the same communication techniques, including tools and software, that your project peers use.

Most technical communication managers are first-level supervisors. Communication among supervisors usually focuses on removing barriers to getting work done and resolving problems associated with the processes that are being used. Other supervisors do not want simply to hear you complain that problems exist. They want you to help solve the problems. Any time you complain to another manager about a problem, your complaint should be accompanied by possible solutions for the problem.

Another common interaction with other managers requires a technical communication manager to complain about SMEs who are either not supplying enough information or who are not reviewing drafts by the dates necessary to meet deadlines. In this case, the technical communication manager must calmly and rationally communicate the nature of the problem, the consequences if the problem is not fixed (missed deadlines), and some ideas for how to change things. A technical communication manager wants to appear to be helping to solve problems rather than simply complaining about them. Scientific and technical employees often complain that technical communicators want to be spoon fed information, that they don't want to learn the technical details about the product or process they are writing about. It is important to anticipate that bias and to have answers for it before communicating with a peer about information access and review schedule problems. The important point here is that you should communicate that you want to work with the other manager, not that you are accusing him or her of having an uncooperative staff. You can always begin and end difficult discussions with reminders that you are all working toward the same goals and that it is important to work together to try to accomplish them.

Communicating with Management

As you move up the management chain, the types of communication change considerably. Upper management gets paid to look at forests and not at individual trees. They want to know how something is affecting overall organizational goals, not about how the minute details are affecting your life. Many first-level supervisors, having recently come from the employee pool, fail to realize that their immediate supervisors and managers higher up want executive summaries and rarely want to hear about details. They also want to hear about what you are doing to help achieve organizational goals, not about all of the problems in the way of doing so.

Immediate Manager

Communicating with one's boss is a complex aspect of any employee's work life. Obviously, the specific personalities of the two people involved will significantly affect the kind of communication that goes on. However, there are a few general ideas that should help any technical communication manager to communicate more effectively with his/her immediate manager.

First, you should do a communication analysis of your manager. Many employees do this in an informal, often unconscious manner. But it is a good idea to analyze in some detail the communication preferences that your boss has and to communicate with him or her accordingly. Look through the lists in this section and at the communication analysis form, and consider each of the variables related to your manager. It is so important to have good communications with your manager that it is critical to understand which methods will work and which ones will not.

Second, remember that your boss is a second-level manager. As such, he or she should be spending some time on strategic planning, budgeting, longer-term goals, and, in general, the "big picture." While all managers should strive to remove impediments to their employees' work, your boss expects you to do so whenever possible, without forcing him/her to get involved. Further, your boss expects you to solve problems, not just to bring them up.

There are two broad categories of problems you can present to your boss, what Anderson (1994) calls "problems of dissatisfaction" and "problems of aspiration." With problems of dissatisfaction, something is wrong that is causing people to be dissatisfied. Your customers may be dissatisfied with the documents they are receiving. Your employees may be dissatisfied with the tools or the procedures they have to use. There may be corporate guidelines or policies that are causing dissatisfaction because they inhibit innovation and improvements in processes. In any of these cases, the solution is to change things so that the people involved move from dissatisfaction to satisfaction. It is important to note that in some cases you have the power (in terms of authority and resources) to do so and that in others you simply do not. In those cases where you do have the power, your boss expects you to solve the problem, or to solve it after consultation. Your communication with your boss, then, about the problem contains four main elements:

1. A definition of the problem and its impact,
2. Proposed solutions to the problem, including the pros and cons for each of the possible solutions,
3. Your recommendation as to which solution to choose, and
4. A request for permission to proceed.

In those cases where you do not have the power, your communication has the same four parts, except that you are also asking your boss to delegate enough resources and authority so that you can solve the problem or to exercise his/her own resources and authority to solve it.

In the case where neither you nor your boss has the power or authority to change the condition—related, perhaps, to organizational policies—you need to alter your communicate approach. Simply complaining about the problem makes you sound like a whiner. It also puts your boss in an awkward position. He or she cannot change the condition, but may not want to defend it, so you cause him/her to have difficulty in responding to you. Your communication approach here should be to acknowledge that neither of you has the authority to change things and that you do not expect your boss to do so. You can propose to do one of two things, either propose to upper management that the policy be changed or decide how you can best live with it. In that case, you are still proposing each alternative you can conceive of, and its accompanying pros and cons, but the outcomes will not be solutions but rather a choice of the least of possible evils.

The second kind of problem, of aspiration, requires yet a different mode of communication. With a problem of aspiration, some desired goal or state does not exist, but you and others want it to exist. Rather than fixing an existing problem, you are trying to create something new. Here, you must communicate the following:

1. The state or condition you hope to create
2. The consequences of not creating that state
3. The costs and resources required to create it
4. A demonstration that the new state will provide benefits that exceed the costs
5. Your recommended method for creating the new condition and a proposed schedule
6. A request for permission to proceed

Managers learn to think about actions in terms of budgets and costs. While you might be inclined to talk about consequences and costs in human terms (my writers are unhappy because they don't have a color laser printer), your boss is much more likely to listen to a proposal if it states consequences in dollars. If it is costing you $20,000 in extra production costs each year to prepare color separations because you do not have a printer on which to test them, you have a stronger argument than one saying that your writers think it would be cool to get color drafts. Your communication would state (1) that you want a color laser printer, (2) that it now costs $20,000 per year because you don't have one, (3) that a new one would cost $10,000 and can be ordered immediately, (4) that the cost-benefits analysis shows that the printer would pay for itself in six months, (5) that you will meet with purchasing to discuss brands and models and will check with the computer network people to ensure that you buy something compatible, and that you can do all of that within the next week, and (6) that you request authority to implement the plan to purchase the printer. Communication about problems of aspiration basically amounts to writing a proposal.

Frequently, technical communication managers report to someone who did not come from the technical communication ranks and who does not fully understand what we do. In that case, your ongoing communication goal is to educate your manager as well and as often as you can. That may require writing longer reports with fuller explanations and justifications than would otherwise be necessary. However, you want to arm the manager with as much knowledge as possible, as that person is going to be representing you in management meetings and budget allocation sessions.

If you are faced with reporting to someone who shows indifference toward technical communication or toward your employees who are technical communicators, your task is even more difficult. How can you be an advocate for employees if you have an indifferent manager? First, you should probably give up trying to persuade that person that technical communication is a wonderful, noble, challenging profession full of top-notch, intelligent thinkers. Instead, you should use the kinds of arguments that are likely to work, which are usually related to the goals and aspirations of the overall organization. Showing how you contribute, along with that manager, toward those goals is more likely to win some grudging acceptance than an argument about the merits of technical communication as a profession or technical communicators as people. Approach this kind of communication task from a manager's point of view rather than from a technical communicator's point of view. Demonstrate the ways in which you contribute to the bottom line, the ways in which you add value (see the Value-Added subsection in the Finance chapter).

Upper Management

Upper managers are usually interested more in outcomes and results than in details of implementation. They assume that you are going to handle those details adequately to achieve the desired outcomes. Likewise with communication, they want to hear about the significant results, not about the specific steps taken to achieve them. It is a good idea in any communication with upper management, be it a presentation, a report, or even an e-mail, to provide an executive summary with the main facts and/or recommendations right up front. If they want the details, they can find them later. But they often do not.

Employees and lower-level managers are often fascinated by the processes they follow and on ways to improve those processes. For many, much of their working time is spent on such an effort. Higher-level managers, though, are not so interested in the fine points of how you are doing things. They want to know about the results and how those results affect progress toward organizational goals. It is a good idea in any discussion with upper management to tie your conversation to larger project, departmental, and organizational objectives, to show how what you are doing fits in the overall picture. This is especially true for technical communicators, who are often doing work that others do not fully understand or appreciate.

Communicating Externally

Technical communication managers must communicate with a surprisingly large group of external parties. Foremost among them should be customers or audience members for the documents that the technical communication organization creates. But the group can also include contractors, recruiters, printers, vendors, media, academics, and government officials.

Communicating with customers is essential to producing high quality documents that meet the needs of our audiences. Many technical communicators pride themselves in having a sufficient level of empathy to accurately assume the role of the audience members themselves. Many of them are wrong. It is especially important during a period when technologies and the ways that people interact with them are changing rapidly that we remain in contact with people who actually use the technologies and concepts about which we write.

At a surprising number of scientific and technical organizations, technical communicators are not encouraged to have contact with customers and, at some of them, are expressly forbidden to do so. This is despite the almost universal agreement among technical communication researchers and authors that such contact is essential. What do we do about this? First, if there is no policy in your organization expressly forbidding such contact, then initiate customer communication without asking. As the old business adage says, it is easier to ask forgiveness after the fact than to ask permission before it. To persuade recalcitrant managers that customer contact is necessary, cite this book, Hackos's 1994 book on project management, Schriver's 1997 book on document design, and practically any technical communication text that has been published in the last 20 years. Explain that quality documents help customers perform the tasks they need to perform rather than merely describing hardware and software systems. Many hardware and software engineers (who are often the managers you are trying to persuade) believe that documentation should "document" the system, that is, should explain how each piece of it works. They need to be educated to the fact that explaining how every component of an automobile works does not teach someone how to drive, which is the primary task they want to accomplish.

If you simply cannot meet with customers, then you need to gather as much information about them as possible. Your two best sources for doing so are your customer relations or hot line group and your sales group. People in those organizations often have detailed information about the audience. Especially helpful are the customer relations group's logs and records about the kinds of problems customers call in about. Even if you have complete access to customers, it is still a good idea to get periodic copies of customer relations' logs and reports, which can often help improve the documentation by showing places where information is not clear enough or has been omitted.

If you are able to meet with customers, consider establishing an ongoing customer group with whom you have quarterly or semiannual meetings to go over documentation concerns. Such meetings need not be expensive; conducting them by way of teleconferences or Web-based meetings can keep the costs down. You can also use telephone surveys, interviews, customer satisfaction surveys, and many other methods to gather information from customers.

One important consideration about communicating with customers is to identify which customer group you are talking to. Depending on the nature of your product or service, you can (and probably do) have multiple customer groups, including those who make purchase decisions (who may not be the users), those who install and maintain the product (who again may not be the users), those who pay for the product (management usually—again perhaps not users), managers of the users, and, of course, the people who actually use the product or service. Talking to only one of these groups can lead to document design decisions that do not meet the needs of one or more of the other groups (Dicks 1996).

Aside from customers, communication with external groups can be so varied as to defy analysis. One essential for technical communication managers to remember is that many discussions with outside vendors, printers, recruiters, and others can, in essence, be contractual. You need to be very careful and very specific about dates, finances, and commitments, and to specify that agreements about them will be final only when put in writing and signed by both parties. While it may seem more comfortable to work with outside groups on a less formal, "word of mouth" basis, it can too easily lead to misunderstandings and missed deadlines later.

Meetings

Technical communication managers have to attend many meetings. We may have to attend the meetings of our own technical communication organization, our own group meetings, project meetings for one or more products or services, and the usual array of meetings devoted to human resources issues, management issues, planning, budgeting, etc. Most of us have had weeks where we spend 20–30 hours or more in meetings, and we have to deal with having multiple meetings scheduled at the same time (five simultaneous meetings is my all-time record).

Many of the meetings in scientific and technical organizations are dreadfully inefficient, for two primary reasons. First, because technical personnel are often not as confident using the written word as we are, they save issues (however petty) for face-to-face meetings rather than working them out via e-mail or other means. Second, because the various groups associated with a scientific or technical project often interact only during a meeting, they try to use the meeting to solve interface conflicts and misunderstandings, even if such conflicts are not the concern of 90 percent of the people in the room. While we may not have much choice but to endure these inefficiencies, we can ensure that our own meetings are run more effectively and that they might perhaps serve as models for other groups as to how to use meeting time better.

Technical communication managers host meetings of many kinds, including:

- Group meetings
- Project status and progress meetings
- Planning and design meetings
- Estimating meetings
- Tabletop review meetings
- Project wrap-up meetings
- Customer meetings

Because a technical communication manager is going to spend so much time in meetings, it is important that we run them effectively (Barchilon 1994). The best way to do so is with improved communications, which should work nicely with our skills and knowledge.

Before calling a meeting, you should first consider whether it is necessary or not. Meetings are expensive. While you are sitting in your next boring meeting, occupy your time by doing some basic math. Consider the loaded salary of each person in the room. If you do not know your organization's loadings onto a base salary, multiply their approximate salary by a conservative loading number like 2.5. Hence, someone making $50,000 would have a loaded salary of about $125,000 per year. Divide 2,000 hours per year (which again is very conservative) into the loaded salary to arrive at an hourly rate. For the $50,000 salary that would come to $62.50 per hour. Multiply that figure by the length of the meeting, say three hours, for a total of $187.50, the cost to have that person at that meeting. If there are ten people in the room with similar salaries, the meeting costs $1,875. If you have the same meeting every week, it costs $97,500 per year. To use a paradox, one of the best ways to have a meeting is not to have it. Ask yourself if there is some more effective and less expensive way to communicate the information or to accomplish the tasks that would be achieved in a meeting.

If you have concluded that you must have a meeting, the next step is to state an objective for the meeting. If you can't do that, you don't need to have the meeting. Because they are such an expensive way to communicate, meetings should not be called if more efficient methods of communication are available. If fulfilling an objective can only be accomplished with a meeting, then it is time to have one. To have effective meetings, you need to take steps before, during, and after the meetings.

Before the Meeting—The Agenda

Prior to every meeting, you should issue an agenda. This is true even for routine, weekly progress meetings. The agenda should contain the following information:

Meeting title or name

Date

Time (including duration)

Place

Attendees list

Meeting objective

Information required (before or during the meeting) and responsible parties

Discussion list, with people responsible for leading discussion and approximate duration

Any other pertinent information

It is important to notify those who are responsible for supplying information before and during the meeting, so that they will have enough time to gather the information and either send it prior to the meeting or bring it along. You should send the agenda several days ahead to give everyone ample time. Also, if the meeting is not a routine one, it is a good idea to send an e-mail reminder the day prior to or the morning of the meeting.

The Meeting Agenda Worksheet provides a model for preparing an effective agenda.

During the Meeting

Dozens of books, articles, and Web sites have covered how to conduct successful meetings. It is not within our scope here to attempt to summarize this vast body of information about meetings, but to discuss briefly the things that a technical communication manager should pay special attention to regarding meetings.

It is important that someone maintain a record of the discussions that occur and the decisions that are reached during a meeting. You can appoint one person to do this permanently or rotate the responsibility. One argument for giving the job to the same person is that the meeting minutes will have a standard level of detail and format, and that there is one person accountable for getting them done and delivered in a timely fashion.

It is important for a meeting facilitator to keep the conversation focused on the agenda items. Allowing forays into tangential areas often extends meeting times dramatically. In a brainstorming meeting, such forays are welcomed and are part of the process. However, in a status meeting, they can take up considerable time. People should be encouraged to solve technical problems outside of meetings so

Technical Communication Management Worksheet

Technical Communication Meeting Agenda

Meeting Title or Name

Date:

Time (including duration; e.g., 9:00-11:00 A.M.):

Place:

Attendees list (Include actual names, if possible):

Meeting objective:

Information required (before or during the meeting) and responsible parties:

Update on XYZ document	Jane Doe
Status of new template	John Doe
Etc.	

Discussion Items (topic, people responsible for leading discussion, and approximate duration):

Introduction	Kim Kilroy	10 min.
Update on XYZ document	Jane Doe	20 min.
Demo of new archive tool	Terry Taylor	20 min.
Status of new template	John Doe	10 min.
Etc.		

Any other pertinent information

that the meeting can follow its agenda and end in a reasonable time. It is also important for a meeting facilitator to be aware of the various types of conversations that occur in meetings.

Types of Conversations at Meetings

Have you every been in one of those long meetings where you have struggled through brainstorming about how to solve a problem, argued and negotiated about the various possible solutions, winnowed those solutions down to two or three choices, argued some more, and finally been headed toward consensus and conclusion, when someone goes back to the brainstorming stage and suggests an entirely new idea? The moans and evil-eyed stares communicate the frustration that the attendees have when someone tries to change the nature of the type of conversation that is going on.

One of the biggest reasons for confusion and waste of time during meetings is that two or three different types of conversations are occurring. According to Begeman (1999), you can have three types of conversations in a meeting:

1. **Conversation for possibility**—This is a conversation where the group is trying to innovate, to generate new ideas, to brainstorm. They are not trying to arrive at decisions or assign actions. It is always interesting to see some people who are so action oriented that they are uncomfortable discussing possibilities. They constantly try to steer the group away from considering new ideas and toward deciding on some action. In the technical communication world, the most common conversations for possibility concern the initial planning and design stages for a new document or set of documents. The group brainstorms about all of the possible types of documents that would meet audience needs for a particular set of information.

2. **Conversation for opportunity**—In an opportunity conversation, the group is trying to winnow down a list of ideas or alternatives, discussing the positives and negatives associated with each. The brainstorming period is over, but the goal is not necessarily to arrive at any conclusions about actions to take. The most frequent technical communication conversations about possibility occur during planning when the group is trying to reduce the possible documentation choices to the most likely ones to succeed. Such conversations also occur when the group is narrowing possible solutions to difficult documentation process problems and to particularly problematic sections or parts of a document. One of the most common problems with meetings in scientific/technical environments is that two people spend an inordinate amount of time arguing about which of two possible solutions is preferable. This often becomes a clash of egos that compels the participants to defend their positions as if they were sacred and holy (for real fireworks, get one started on the Mac OS versus Windows!). Some project meeting coordinators have a standing rule that solutions will not be argued during a meeting but that the parties involved in arriving at a solution will be given an action item and must resolve it outside of meeting time. This is a good policy because otherwise you can spend way too much time in

meetings where two people in the room fight while the others wait impatiently for it to be over. See the earlier analysis on meeting costs for the main reason why we should avoid letting this happen.

3. **Conversation for action**—Here, the group arrives at conclusions and assigns action items. The purpose in this type of conversation is to leave the meeting with firm commitments and assignments for completing tasks. The most common conversations for action for technical communicators occur in project status meetings, where, after discussion, problems and their solutions are assigned to specific people with specific deadlines.

By far, technical communicators spend more time in project status meetings than in any other type. It is important to note that all three types of conversations might occur during a project meeting. Someone brings up a problem that needs to be solved. The group first has a conversation for possibility, with various people offering ideas for solutions. The group then narrows the list by conducting a conversation for opportunity, comparing the various solutions and moving toward the one that seems most compelling. Finally, the group reaches a conversation for action, where they decide what will be done, who should do it, and by when. The transitions between these types of conversations are usually not stated explicitly by the person chairing the meeting; they seem to occur naturally. However, when they do not occur naturally, when different people in the meeting are having different types of conversations, you can waste an enormous amount of time.

One of a meeting facilitator's best skills is to realize what type of conversation should be going on and to direct people toward continuing that type of discussion. It is even a good idea to state explicitly when you are moving from one type of conversation to another, so that people will stop brainstorming and start narrowing the possibilities. Some people are much more comfortable thinking conceptually and will keep trying to get the conversation back to one of possibilities. Others who are more analytical will be eager to move to a discussion of opportunities, where they have a chance to find the faults and the advantages of various solutions. Those who are analytical to a fault will prefer to criticize the winning possibilities rather than move toward resolution and action. And others will constantly push toward resolution and action. A good meeting facilitator must ride herd over the various preferences for conversation and must ensure that the group spends sufficient time having each type of conversation.

Of course, many other types of conversations can go on in a meeting. In many other cultures, people prefer to begin meetings with extended periods of social interaction, discussing families, histories, the weather, etc., in order to get to know each other before trying to arrive at business conclusions together. They find Americans to be rude and abrupt in our concern to get to decisions quickly. In our meetings, we should include time for conversations that are not necessarily oriented toward any of the three areas discussed above. We might want to allow some social time, some time for questions and explanations, some time for rumors (and debunking thereof), and other matters. For technical communicators, these discussions will usually occur during general group meetings rather than project meetings, which are, by definition, aimed at reporting status and solving problems.

After the Meeting—The Minutes

Most technical communication meetings concern projects, their progress, and the status of the development efforts versus schedules and budgets. Therefore, it is critical that our meetings have outcomes that help everyone stay on schedule. One of those outcomes should be a set of minutes that details all action items identified during the meeting, who is responsible for each, and the dates by which each must be done.

Meeting minutes provide several benefits. First, they act as a record of the progress of a project, which can become important later for CYA (cover your assets) and even legal purposes. Second, they keep everyone informed as to who is responsible for each action item in the coming weeks to ensure that the project stays on track. Third, they act as a communication device, not only for meeting attendees but for others who do not attend but whom you want to keep apprised of your status. Finally, they act as a motivating tool to keep people focused and on track in fulfilling necessary project milestones, including helping people to identify priorities so that they work on the most essential things first.

Because minutes are so important for technical communication groups and projects, you should try to get them issued as soon after a meeting as possible. Some people prefer to try to capture the high points of each discussion that goes on during a meeting, concluding with any action items associated with the discussion. This leads to very detailed minutes that are difficult for one person to keep up with during the meeting, often making it difficult for that person to participate. Others prefer to provide only the essence of each discussion and the conclusions reached. For most technical communication meetings, briefer descriptions of the discussions should be sufficient. For projects that are larger and/or more controversial, more detailed minutes may be required.

Minutes give technical communicators another opportunity to educate coworkers and peers in other groups about what we do and how we do it. Whether you or someone else does the minutes, you should not lose the opportunity to educate people to the fact that we have professional procedures and processes we follow, just as they do. In other words, consider your minutes to be part of the ongoing educational and public relations that you conduct on behalf of the technical communication group. If it is appropriate, send copies to management in other departments, even if they do not attend the meetings.

The Technical Communication Meeting Minutes Form provides an example of what a set of minutes should include.

Note on the Meeting Minutes Worksheet that the discussion section is in chronological order, following the order in which issues were discussed during the meeting. Note also that action items are set off and are put in bold so that readers can easily find them. In addition, the action items are listed in a table at the end of the minutes, where they are assigned a priority. This table can help communicators decide where to spend their time first. The table can also communicate project

Technical Communication Management Worksheet

Technical Communication Meeting Minutes

Minutes: Meeting Title or Name

Date:

Time (including duration; e.g., 9:00-11:00 A.M.):

Place:

Attendees List (include names of those present and those absent):

Meeting Objective:

Discussion Items (topic, people responsible for leading discussion, and approximate duration):

Introduction   Kim Kilroy   10 min.
Kilroy introduced the meeting topic, requested any corrections to the last set of minutes, and introduced J. Doe for the first discussion.

Update on XYZ document   Jane Doe   20 min.
Doe described the problems with creating the XYZ document, which is one week behind schedule. The document is late due to difficulty getting specifications from one SME.

Action Item: Kilroy and Doe will work with Frame expert to get document formatted by June 30th.

Demo of new archive tool   Terry Taylor   20 min.
Taylor showed how the new tool will work and will help us keep better archives.

Status of new template   Joey Dokes   10 min.
Dokes explained that the new template is almost complete and that it should be ready by the July 10th deadline.

Action Item: Dokes to have new template online by July 10th.

Etc.

Project XYZ Documentation Action Items

Action	Responsible	Date	Priority	Status
Expedite receipt of final specs from XYZ SME	Kilroy, Doe	June 30th	High	Started
Put new doc. template online	Dokes	July 10th	Medium	On schedule

progress to external groups. The table is also used at the beginning of each meeting to go over the status of all of the action items, which tends to help people stay focused on the actions required of them.

Meeting Improvement

One of the ways to improve the quality of your meetings is to conduct a brief analysis at the end of each meeting. Spend five minutes listing the things you did well and the things that need improvement. This may be delicate, as the areas for improvement may involve someone's presentation mode or behavior, and that includes you, the meeting facilitator. Include your lists in the minutes so that everyone will see them again and will work on doing more of the successful things and on avoiding those that were not so successful.

References

Barchilon, M. G. 1994. Technical communication models that ensure productive meetings. *Publications management: Essays for technical communicators.* Edited by O. J. Allen and L. H. Deming, 25–38. Amityville, NY: Baywood.

Beck, C. E. 1999. *Managerial communication: Bridging theory and practice.* Paramus, NJ: Prentice Hall.

Begeman, M. 1999. You have to start meeting like this! *Fast Company*: April, 204–210.

DeGraw, Y. 1993. Us vs. them—why can't we be friends? *Technical Communication* 40(1): 80.

Hackos, J. T. 1990. Managing creative people. *Technical Communication* 37(4): 375–380.

Hackos, J. 1994. *Managing your documentation projects.* New York: John Wiley & Sons, Inc.

McLuhan, M. 1965. *Understanding media: The extensions of man.* New York: McGraw-Hill.

McGuire, G. 1991. "Working in 'flow': A theory of managing technical writers to peak performance." *STC 38th Annual Conference Proceedings*, Society for Technical Communication.

Schriver, K. A. 1997. *Dynamics in document design.* New York: John Wiley & Sons, Inc.

Trevino, L. K., R. L. Daft, et al. 1990. Understanding manager's media choices: a symbolic interactionist perspective. *Organizations and communication technology.* Edited by J. Fulk and C. Steinfield, 71–94. Newbury Park: Sage Publications.

Weiss, E. H. 2002. Egoless writing: Improving quality by replacing artistic impulse with engineering discipline. *Journal of Computer Documentation* 26(1): 3–10.

For Further Information

Charles Beck's *Managerial Communication: Bridging Theory and Practice* (1999) offers communications models of increasing complexity as he discusses each of the various types of communication, audiences, and situations. This book provides a good overall study of communication theories and practices for technical communicators.

Trevino, Daft, and Lengel discuss the implications of managers' communication choices in "Understanding Manager's Media Choices: A Symbolic Interactionist Perspective" (1990). Of course, one of the classics related to media choices is McLuhan's *Understanding Media: The Extensions of Man* (1965).

Gene McGuire's article, "Working In 'Flow': A Theory Of Managing Technical Writers To Peak Performance" (1991) discusses the necessity of communicating with each employee according to his/her particular communication preferences. JoAnn Hackos, in "Managing Creative People" (1990), also suggests tailoring communication to individual employees and recommends determining their communication preferences through standardized personality testing.

Marian G. Barchilon covers meeting methods for technical communicators, including sample agendas and minutes in, "Technical Communication Models That Ensure Productive Meetings" (1994). Michael Begemen examines the kinds of discussions that can go on in meetings in "You Have to Start Meeting Like This!" (1999).

Questions for Discussion

1. What are your own communication preferences? Would you rather discuss sensitive issues in person, via telephone, via e-mail? Why?
2. When someone summons you to come to him/her to communicate about something, how does it make you feel? Why?
3. With all of the electronic communication means available to us, why should we bother to have face-to-face communications? Aren't face-to-face communications a passé, inefficient, overly romanticized, antiquated method for conveying information? Why don't we simply do it all more quickly and efficiently via electronics?
4. For one day in the next week, count the number of times you have to change rhetorical situations throughout the day. You will have to carry a tally sheet that is easily accessible because there may be periods where the changes happen frequently. How many did you come up with?
5. Are the meetings in your organization productive? Are they held on a regular basis? Is an agenda sent out in advance? Are action items assigned, recorded, and tracked? Are minutes issued?
6. What experience have you had with meetings that were especially effective? What made them effective? What about meetings that were especially ineffective?
7. Some management gurus contend that organizations waste as much as 20–30% of their employees' time in ineffective meetings. How can you avoid this? Are there other ways of communicating information across several departments? One company went so far as to remove the chairs from their conference rooms, thus requiring everyone to stand throughout meetings. What other methods can you conceive to reduce meeting time yet ensure effective communications?

8. With all of the technology available (e-mail, listservs, newsgroups, group-ware, etc.) to replace meetings, why do we continue to have so many of them? What social and cultural purposes do they fulfill?

9. Many organizations expect managers to use meetings to promulgate the organization mission and goals. They further expect managers to present various new policies related to administration, human resources, benefits, work procedures, and so on. What would you do if you were required to present such information to your employees even if you did not agree with its content?

CASE 4

Meeting Agenda, Minutes, and Action Items

The Management Situation

Aardvark Enterprises has decided to enlarge its technical communication group, with you as the new group manager. Congratulations on your promotion!

You will be required to hold regularly scheduled progress meetings on the Midas project. (See Assignment 1.) These meetings will include your staff members who are working on Midas plus representatives from the system engineering, programming, system test, human factors, customer support, training, and project management groups. It is imperative that you distribute an agenda prior to each meeting and a set of minutes after each meeting.

Your meeting documents will serve two primary purposes. First, they will serve as a history and a repository of information about the development process for your part of the project. Second, they will serve to communicate to your staff and to the other departments what your group is working on, what requirements you have from other people, and how well you are staying on schedule.

The other departments in Aardvark are functioning at level 2 to level 3 on Hackos's (1994) process maturity model. They will, therefore, expect you to hold regular progress meetings and to report the results. They will be watching to see how well you manage the meetings and how well you communicate to them information about your progress and about what they are required to do as part of your development process. Likewise, your own management has made it clear to you that you must develop a regular process for having meetings and communicating the results.

The Assignment

Prepare an agenda for your first technical communication progress meeting on the Aardvark Midas project. This does not have to be a complex document, but it should include, as a minimum, the following: the date, time, duration, and place of the meeting; the list of invitees (plus other recipients who aren't necessarily invited); a list of the topics to be discussed, who is responsible for leading each discussion, and the approximate time to be spent on each topic, and any information or documents that people need to bring to the meeting.

Prepare an action item tracking list that you will use at the meeting to record action items as they come up. The list should include the action item, who is responsible for doing it, the date it should be started and completed, a priority for the item, and current status. This list will also be distributed after each meeting as part of the minutes.

Prepare a set of minutes for your first meeting, including at least five action items. (You will have to create those items and some text for the minutes. The content is less important here than the process.) The minutes should include, as a minimum, the date, time, duration, and place of the meeting; a list of the attendees (including a list of absentees); a list of topics discussed and the resolution of each, noting in some conspicuous manner where action items were assigned; and the date, time, duration, and place for the next meeting.

Helpful Hints

Try some of the following sources for more information about meetings:

Barchilon, Marion G. "Technical Communication Models That Ensure Productive Meetings" in *Publications Management Essays for Professional Communicators*. Baywood, 1994.

Sources cited in the above article.

Begeman, M. You Have to Start Meeting Like This! *Fast Company*, April, 1999, 204–210.

Evaluation Criteria for Case 4

Does your agenda include all of the necessary parts? Does it include any additional information that needs to be added for this particular meeting? Does it provide meeting attendees with a clear idea of what will be discussed and what each of them is required to bring to the meeting?

Does your action item list communicate clearly to everyone which action items are current, which are overdue, and which will be coming in the future? Does it provide information about the consequences of missed dates?

Do your minutes accurately summarize what happened at the meeting and provide all parties with the information they need to perform any action items that were assigned to them? Do the minutes help them to start preparing for the next meeting?

CHAPTER 5

Management Training

Introduction

In many scientific and technical organizations, there is only one group of technical communicators, with one technical communication manager. That means that the technical communication manager reports to someone who likely did not come from our discipline and who may not be able to offer much mentoring or advice about career advancement and training. Hence, the burden often falls on technical communication managers to devise and implement their own training and development systems.

There are five types of knowledge and skills that a technical communication manager should have. All five are important, so managers must do self-assessments and then create their own development schedules, taking into consideration the restraints of budget and of time away from work. The five categories are listed in Table 5.1.

General Management Skills

Most technical communication managers come into their jobs with very little formal training in basic management skills (Hackos 1989; Anderson 1994). These skills are so important that a new manager needs to develop them quickly, and even more experienced managers need occasional "refreshers" or perhaps retraining if technology has changed how the skills are best implemented. Because these skills are general, they can be learned through general training courses, books, online tutorials, and other sources, rather than requiring industry- or organization-specific training courses. One good place to start is the American Management

TABLE 5.1	Technical Communication Management Training Knowledge/Skills
General Skill	**Specific Areas of Expertise**
General Management Skills	Project Management *Eng 625*
	Time Management *Franklin-Covey*
	Personnel Management
	Financial Management *Budget, direct & indirect cost*
	Strategic Planning *goal setting & how to get to goal - incremental series of goals*
	Problem Solving *. think on feet*
Tech Comm-Specific Management Skills	Document Management *- mg update process*
	Information Architecture and Management *- Structure doc, program*
	Document Assessment
	Design of Document Libraries and Individual Documents *- Archive System*
	Editing *- written word effective*
	Estimating *- get handle on cost*
General Technical Skills	Computers/Networks/Communications
	On-line and Web-based technologies
	Distance learning and delivery
	CD-ROM/DVD production and delivery
TC-Specific Technical Skills	Publication development methods and software
	On-line system development methods and software
	html and Web-based development and delivery
Industry- or Discipline-Specific Skills	Technology involved in the general industry of the overall organization (i. e., telecommunications, computer software, pharmaceuticals, etc.)

Association (http://www.amanet.org), which offers a wide range of training materials in all aspects of management. Another good resource, especially related to time and project management, is the Franklin/Covey series of books and training classes (http://www.franklincovey.com).

Project Management

All technical communication managers need to know the basics of project management, as well as how to apply those methods specifically to technical communication projects. JoAnn Hackos's 1994 book, *Managing Your Documentation Projects*, provides a superb discussion of how to institute project management techniques that lead to the highest quality documents possible. Every technical communication manager should read this book.

Depending on the environment, technical communication managers may also need to learn more of the fundamentals of project management, including how to use the tools that are on the market. The American Management Association offers an excellent training course in project management, and there are also many good books and seminars on the market. While Microsoft Project is the "default" project management program that many use, there are many other such systems, designed to run on computer platforms of various sizes. Additionally, an increasing number of Web-based applications for project management are coming onto the market. A technical communication manager should adopt whatever project management tools are commonly used in the industry or discipline in which he/she is working. In the absence of a default, get a copy of Microsoft Project and learn it along with a book or training course. It is simply a basic, essential skill for a technical communication manager to be able to create a Gantt chart showing key project milestones and critical paths.

Time Management

Likewise, time management is crucial to the success of a technical communication manager. Many communication managers supervise people who are working on multiple projects, often dozens at a time across an entire group. This means that a communication manager may have a much larger number of interactions with people in other departments and externally than do most managers. With e-mail, telephone calls, and meetings associated with so many projects, it is essential that a manager keep track of his/her schedule, commitments, and communication requirements. See Chapter 7 for more information on time management.

Personnel Management

For personnel management, Chapter 2 of this book provides basic information. The market is flooded with general books on personnel management, so it is difficult to recommend only one or two. Technical communication managers should belong to the Society for Technical Communication and receive its two journals. One, *Intercom*, often has articles directly related to personnel management for technical communication managers, as well as articles on many other aspects of management. Another good resource is the STC on-line publications database (http://stc.org/search_pubs.html), which provides searches in *Intercom*, STC Conference Proceedings, and dissertation abstracts.

Financial Management

The market is also full of books on general finance issues for managers. One of the better ones for technical communication managers is Finkler's (1992) *Finance & Accounting for Nonfinancial Managers*. While it is important to understand the general basics of finance, it may also be important to learn how finances and reporting function in the particular organization where you work. In that case, you will need to take internal training courses or to get the information from your manager or other mentor.

Strategic Planning

Strategic planning is also covered in many general business books. The November, 1997, issue of STC's *Technical Communication* was a special one devoted to strategic planning for technical communicators. The issue contains several valuable articles and case studies on strategic planning, especially Hackos's "Using the Information Process-Maturity Model as a Tool for Strategic Planning" (Hackos 1997).

Problem Solving

As Anderson (1994) points out, another important set of general management skills for technical communicators involves learning how to adopt an action-oriented approach to problem solving. Technical communicators rarely receive training or instruction in solving problems, and yet doing so is one of the most fundamental management skills. Anderson suggests a basic five-step, problem-solving process:

1. Define the problem
2. Design the solution
3. Test the solution
4. Implement the solution
5. Evaluate the solution

While these steps may seem quite obvious, it is surprising how many people cannot or do not apply them when trying to solve a problem. While many technical communicators may in fact do these things subconsciously while they are writing, they have often not concentrated on problem solving as a skill in itself. Hence, when they are confronted as managers with difficult problems, they may take an ad hoc, unstructured approach to solving them. Technical communication managers need to work at solving all types of problems associated with their function, not just those associated with getting good documents out. Those problems may involve related areas where the manager does not have much expertise, such as network problems, computer problems, software problems, process problems, and others. It is especially helpful, when confronted with new problems where one does not already know the answer, to have a standard methodology for dealing with the problems. The system that Anderson offers provides a good start in developing problem-solving skills.

It is doubtful that one can learn enough about general management skills simply by taking the occasional seminar or training class. To some extent, learning about general management concepts and skills needs to become an ongoing part of one's professional life and, to an extent, private life. One way to constantly improve general skills is to watch how respected managers do things and to ask them about how and why they do them. Another way is to subscribe to business magazines such as *Fortune*, *Forbes*, and *Fast Company*. One does not master general management skills in a year or two. Some of the time spent learning and improving such skills will almost certainly have to come outside of normal working hours as one reads books, magazines, and journals devoted to management.

Technical Communication-Specific Management Skills

Aside from the general management skills that all managers must learn and improve constantly, technical communication managers must also master a set of skills related specifically to the field. These special skills depend on the type of industry within which one works, so that not all technical communication managers must learn all of the skills, but there are some that almost everyone must know, such as document management. Other special skills include information architecture and management, document assessment, editing, and estimating.

Document Management

Even in the days of paper documents, managing versions, masters, and drafts of a single document could be a nightmare. With much documentation now developed and delivered strictly on-line, the problems become even more complex. How do we prevent the all too familiar phenomenon of multiple "master" copies of a document, so that we have to spend time reconciling them or, worse, that we send the wrong one to customers? How do we ensure that we have master files available for previous versions of our products in case we should need to print more documents or deliver the on-line version to someone whose disk drives have crashed? How do we keep all of the paperwork and files associated with a submission to a government agency? A technical communication manager is eventually going to have to deal with these questions, and it is far preferable to deal with them in advance rather than after some crisis has occurred. Documentation management is a skillset mandated by the nature of the work we do.

At too many organizations, the document "control" system consists of each communicator keeping master files on his/her hard drive, often without any backup system at all. Each person names the files according to his/her own preferences. When someone leaves or gets sick, the others have to wade through thousands of files to try to determine which ones they need to work on. Worse, when the hard disk crashes, as it eventually will, days or weeks of work may be lost.

Fortunately, there are several turnkey documentation management programs available on the market. However, they are generally expensive, difficult to install and maintain, and often more complex than a small technical communication group might need. Some of the packages are designed generically and are intended for use by anyone, while some of them are designed for specific, vertical markets, such as pharmaceutical submission systems or physician compliance systems. Vendors for such programs tend to come and go quickly, although Xerox and Documentum have been in the field for many years. You can sometimes find ads and training courses related to document management advertised in STC's *Technical Communication* and *Intercom*. Also, searching on the Web for "document management," "documentation management," and "content management" will yield additional possibilities.

If your technical communication operation is too small to warrant one of the commercial solutions, you will need to come up with one yourself. A good document control system should include, at a minimum, the following:

- File naming conventions
- Secure master files, including graphics
- Version control
- Off-site storage
- Daily and weekly backups

The purpose of the document management system is to ensure that you have all versions of a document that a customer might need, including those that are several years old, as some customers may choose not to upgrade software or hardware even though newer versions are available. Further, the system should make it easy to find files when their "owners" are not around. It should also ensure that master files are kept separately somewhere so that they cannot be inadvertently modified, and that such files should be stored off-site so that in case of a catastrophe you will still be able to function. Finally, the system should ensure that you have backups that preclude having to re-create documents or on-line systems should hardware or software glitches cause the work files to be lost.

Information Architecture and Management

"Information architecture" is used to mean many things, depending on the nature of one's job. Programmers, librarians, technical communicators, and Web site managers all tend to think about it differently. For the purpose of this discussion, information architecture relates to the physical structure of a set of information and to the media used to communicate it. One of the most important and difficult skills for a technical communicator comes after assessing the tasks and needs of a specific audience to whom a particular set of information must be conveyed. The communicator must first determine which of many document types will best accomplish the task. This may require one document type or several, depending on the communication needs of each audience set. For a piece of software, for example, we might decide that installers need a paper guide, users need an on-line help system, administrators need a paper admin guide, and data entry personnel need job aids. Meeting customers' communication needs requires that we design the entire system's architecture so that everyone can accomplish the tasks he/she needs to accomplish.

The communicator must then determine what media will be used to deliver the information. If the information will be housed electronically, we must determine whether it will be in a database, in disparate text and graphics files, in folders on the local hard drive, on a Web site, etc. Not only must we design the information set to meet customer needs, but we must also deliver it via media that work with the customer's daily environmental and task requirements. Obviously, for example, having only Web-based instructions for someone who is installing an antenna at the top of a 300-foot tower is not an acceptable solution.

A technical communication manager should know how information architecture works for several reasons. Even if you rely on your employees to design document solutions, you should be able to discuss them intelligently and to challenge their decisions. Further, you may have to defend the architecture decisions to others, so you will have to understand how they were arrived at. And, you may have to balance the optimal design solutions that your employees develop against the realities of time and budget constraints.

Most technical communication courses discuss information architecture as it relates to developing a single document, but they often do not discuss in much detail the overall product or service or the entire system architecture of the information set to be delivered. To some extent, learning this skill set is a function of experience. You can learn more, though, by carefully searching for courses, undergraduate and graduate, in technical communication, computer science, library (or information) science, and communications offerings. In addition, there are a number of seminars and classes offered regularly, although you should read the descriptions carefully to ensure that they are treating information architecture in a way that will be helpful to you, rather than looking at it strictly from, say, a programming point of view.

"Information management" usually refers to the manner in which the information will be stored and accessed by the audience. Will it be on a Web site in separate files? Will it be in a database? If so, should the database be flat or relational? How will the interface to the data work? What will customers see and do to get the information they need? The most well designed information set is rendered worthless when the means for getting to it are awkward or indiscernible. Accordingly, a technical communication manager should also learn about the field of information management, which is constantly changing as new and more sophisticated technologies develop. While you can take a course to learn the basic concepts of information management, you should also plan on constant reading to keep up with the changes. To some extent you can do this by reading STC publications, but it is also a good idea to read computer-related journals that discuss information management concepts and tools, such as *PC Week* (on-line) and *Infoworld*. Another good source with links to many information/knowledge management resources can be found at the government-maintained Web site at http://www.km.gov/index.html. Also valuable is the Knowledge Management Consortium International's Web site at http://www.kmci.org/.

Document Assessment

All technical communication managers with supervisory responsibility will have to assess documents. You will have to determine how well your employees are meeting the needs of your customers, taking into consideration the time constraints under which those employees are working. You will also have to decide which criteria you apply to your assessments. Is customer satisfaction the most important criterion? How will you know if someone has achieved it? Is meeting industry standards or corporate standards more important? Is technical accuracy most important? And what about meeting your own internal standards for style, grammar, punctuation, spelling, etc.? It is very easy to set up a document assessment system that measures something and that seems to provide a basis for comparison among a group of writers. However, the criteria you use will affect the results you get. If you choose technical accuracy, for example, the communicators will spend their time making sure the documents are accurate even if they don't really meet the needs of the audience. It is important, then, for technical communication managers to choose document assessment criteria carefully.

Because analyzing documents prepared by your employees is such an important skill, it is one that managers should develop. Many technical communication courses and seminars have a segment devoted to assessing documents. Many of the textbooks include examples of document quality checklists. Such checklists often lend themselves to examining the mechanics of a document more than the overall quality of it in terms of meeting audience needs. Nonetheless, they provide a starting place for learning more about the various kinds of document assessment and a means of ensuring that you do not overlook anything significant. Another activity that can improve document assessment skills is participation in judging STC competitions. If your local STC chapter sponsors a competition, volunteer to be one of the judges. You will receive a few hours of training and will work with others to judge a set of documents, which can give you valuable insights as to how other people assess documents. Another technique that can yield interesting results is to sit down with all of your employees to assess an external document, perhaps, for example, a competitor's document for products or services similar to those you work with. The group can often learn much from assessing which things the document does well and which it does poorly. Such a session can also give you a forum for communicating to your employees those aspects of a document that you consider to be the more important ones.

Editing

Some larger technical communication organizations have a staff of editors who work either as a pool, taking editorial assignments as they come in, or as individuals assigned to specific projects. Frequently these editors first perform comprehensive edits on plans and early drafts and then perform copyedits and approvals on final drafts. They also often serve as general editors for a project, ensuring consistency among all of the project's documents.

In smaller organizations, however, there may be only one editor or, perhaps, none. A technical communication manager must decide how editing will be handled in such an environment. One common practice is to require communicators to edit their own work and to carry full responsibility for accuracy and adherence to the organization's style and quality requirements. This sounds good, but even the best communicators are often very poor at editing their own work. In some organizations, the communication group manager serves as the editor for everyone. This, of course, requires that the manager have editing skills that match the types of edits that must be performed. If those are substantive edits, then the manager must know how to look at the overall organization, structure, and style of a set of documentation and an individual document. If copyedits are required, the manager must be able to read documents closely to discover misspellings (and no, spell checkers do not find them all), grammar, and style problems.

Learning and improving editing skills is largely a function of doing edits on a repeated basis. However, there are some valuable workshops available through organizations that advertise in STC's *Technical Communication* and *Intercom*. Additionally, many university undergraduate and graduate programs in technical communication include at least one course in editing.

There are also several good books on the subject, including Bush & Campbell's (1995) *How to Edit Technical Documents*, Samson's (1993) *Editing Technical Writing*, and Eisenberg's (1992) *Guide to Technical Editing: Discussion, Dictionary, and Exercises*.

Estimating

As discussed in the Finance chapter, estimating is a critical skill for technical communication managers. Several of the technical communication training companies offer courses or parts of courses in estimation. Check the advertisements in *Technical Communication* and *Intercom*. See also Hackos's (1994) chapter on estimating. Based on the system discussed in that chapter, spreadsheet estimating templates are available from Comtech, Inc. at http://www.comtech-serv.com/.

General Technical Skills

Because almost all technical communicators work on computers, it is essential that a technical communication manager learn something about computers and how he/she works. Obviously, those managers who work in the computer-related hardware and software industries may need to learn much more. But every manager should understand how their group's computer system works, how a network works, how the Web works, and how the various media (CD, audio, video) for distributing information work. It is important for a manager to know what technical challenges his/her employees face and to help smooth the way so that they can work as effectively as possible. With no understanding of how the systems he/she uses work, a manager faces a much more difficult job. Further, many employees assume that a manager should understand the technology they use and should be open to suggestions for upgrades in equipment and software. A manager who evinces no understanding of the technology risks losing credibility in the employees' eyes.

In some smaller organizations, the technical communication manager may in fact become responsible for running the group's computer system, for getting machines maintained, repaired, and updated, for handling glitches that occur, and for purchasing upgrades and new products. Further, with information increasingly being delivered via media other than paper, a technical communication manager should understand what the alternative media are, what the strengths and weaknesses of the various media are, and what technology is necessary to develop and deliver documentation using each medium.

How does a manager keep up with all of this? Books and college courses can help, but the lead times necessary for them often mean that they are not up to date with the latest technologies. Training courses from vendors in our field and in related fields can help. Again, look at the companies that advertise in *Technical Communication* and *Intercom*. Most effective, though, because of their shorter lead times, are magazines and Web sites. Magazines such as *PC Magazine*, *PC World*, *Computer Shopper*, and others can help a manager to stay current with developments in computer hardware and software. *Infoworld* and *PC Week* (on-line) are more oriented toward computer industry news, but they also provide information about the underlying technologies that are being developed and introduced. The ZDNET Web site (http://www.zdnet.com) provides current news about technological developments. It is a good idea to take at least one of the paper magazines and to read one of the Web sites regularly.

There are many training and tutorial methods available for improving computer and network knowledge. Several of the training companies listed in Chapter 9 offer training not only in technical communication related subjects but also in computer operating systems, networks, and applications. See their Web sites for more information.

In addition, you may need to keep up with more specific technologies. If your organization, for example, is developing products that will function on personal digital assistants (PDAs), you might need to subscribe to one of the PDA magazines and/or to consult with one of the Web sites devoted to news about that technology.

Technical Communication-Specific Technical Skills

Aside from the general technical skills that any technical communication manager should have, there is an additional set of knowledge and skills that are particular to the technical communication field. This includes areas such as publication development methods and software, on-line system development methods and software, and html- and Web-based development and delivery.

Publication Development Methods and Software

One of the necessary jobs for a technical communication manager is to stay abreast of the developments in methods and tools for creating documentation in several media. This means keeping up with the various platforms on which documents can be prepared, the file formats they can use, and the software products that work on those platforms using the various formats.

Platforms—Are you a Mac shop or a Windows shop or a UNIX shop? To a large extent, this will be determined by what everyone else in your organization uses. It is important to be able to exchange files easily with your SMEs and reviewers, so you probably want to be on the same platform as they are. Reading the computer journals such as *PC Magazine*, *PC World*, *Infoworld*, etc. will help you keep up with news about developments in platforms.

You also want to keep up with the software tools available for creating documents. While Adobe Framemaker is the current tool of choice for preparing large, paper documents, this could change as more and more documentation is delivered on-line rather than on paper. Framemaker has the advantage of having many features that assist and automate the development of books. The other most commonly used program, Microsoft Word, works fine with smaller-sized documents (up to about 50 pages), but its weaker book-oriented features make it less effective with larger documents. Word, however, has the advantage that nearly everyone knows how to use it, which makes it easier to hire people fast when you need to ramp up. When it comes to tool selection, one of the constant tradeoffs is between the tool that seems to fit your needs best and the tool that is used by the most people. While the one that fits better might make your group more productive and give you capabilities that you don't have now, when you need to add staff you may have a difficult time finding new employees and contractors who know how to use it. On the other hand, using the more popular tools makes staffing easier but may not meet all of your development needs. While almost anyone who has used one word processing tool can use another for simple tasks, to get someone to the point of offhand ease using such tools can take months and even years, during which time their productivity will be lower than it will be later. Hence, ease of learning and use is yet another consideration when selecting tools.

Keeping up with which tools are standards and which new ones are available requires not only reading the computer magazines, but also looking through literature that is specific to the technical communication field. You can look at advertisements in the technical communication journals and see which vendors appear at conferences. The STC conference is an excellent place to get an idea of the range of tools available including, often, demonstration copies of them. Another place to look is in the classified ads for technical communicators in newspapers and on Web sites. The ads often specify the tool knowledge that candidates should have. Looking through these ads will give you an idea of which tools other technical communication groups are using.

On-line System Development Methods and Software

The same methods apply for keeping up with methods and tools available for creating on-line documentation and help systems. Here, however, there are many more tools and no front-runners in terms of usage. While RoboHelp is a very popular tool for developing Windows- and html-based help systems, there are many other programs on the market. With the distinctions between dedicated help systems and Internet-based help systems blurring, keeping up with the available tools requires constant vigilance through perusing the journals, magazines, and Web sites.

html and Web-based Development and Delivery

With the Web increasingly becoming the preferred platform for delivering many kinds of information, an entirely new set of methodologies and tools has become available. There are dozens of Web site development tools, graphics packages, 3-dimensional rendering products, video editors, audio editors, etc., available to automate Web site development. Keeping up with all of them is almost impossible. Again, you can stay current with the computer magazines and journals, the technical communication journals, and the Web sites. Attending conferences to hear about what others are using is also helpful. Following the mailing lists for technical communication will give you insight into what people are using, based on the types of questions and problems they are reporting. Because this field is changing so rapidly, the companies developing and offering tools come and go quickly. While the word processing developers with large existing user groups are likely to stay in business, many of the smaller firms using newer development technologies might not have so secure a future. Before investing too heavily in purchasing a tool and training employees in its use, you should check out the financial viability of the company that is selling the tool. It is expensive and time-consuming to port software from an abandoned program over to another one or, as is often the case, to have to develop it from scratch in another program.

Should a Manager Know the Tools Better than the Employees?

Most technical communication managers get promoted because they develop excellent documents, which means that they know how to use the tools very well. In some unfortunate cases, the best tool jockey gets promoted whether he/she has any knowledge about what good documentation is or the faintest notion as to how to manage. Up until a few years ago it was possible for a manager to know the tools as well as or better than the communicators doing the development work. Because the manager usually came from that group, he/she was already familiar with the tools. Hence, the manager could serve as mentor and coach to the employees, helping them when they had questions and problems using the development technology. Until recently, most technical communication groups used a single word processing package and perhaps, one graphics package. So, it was reasonable to expect that the manager of the group would also learn how to use those tools.

Because of the proliferation of platforms, methods, and tools available to technical communicators, it has become extremely difficult for many managers to know every tool that everyone in the group is using. One manager might have people preparing paper documents (one set of tools), on-line help systems (another set), on-line tutorials (another set), Web sites (another set), and CD-ROM/CD-RW/DVD training packages including video and audio (several more sets). For a manager, who must also spend considerable time on personnel, budget, and other concerns, to learn all of the tools involved would simply be impossible. What strategy can you adopt if you have too many tools to learn? You can try to work through basic tutorials for each tool so that you can talk intelligently to the person who is using it and so that you will have reasonable expectations about what it can and cannot do. If there are too many even for that, you can have each developer give you a brief tour of the tool and explain how it is used to accomplish their various tasks. While you may not be able to be an expert mentor for every tool, you want to support the staff members using them in every way possible, including paying for training and how-to books that will help them learn the tools as quickly as possible.

If you are in an environment where only one or two tools are used, you should try to learn them and to become an expert in their use. If, however, you are in an environment where multiple tools are used and it is impossible for one person to know them all, then you should try to become at least somewhat knowledgeable about each of them and to assist your employees in learning them in any way possible.

Industry- or Discipline-Specific Technical Skills

Technical communication is technical (Dobrin 1983). Most managers of TC groups work in scientific or technological environments, usually specializing in one particular discipline or industry. How important is it for a technical communication manager to learn and understand the underlying science or technology of the products or services that his/her employees work on? Some argue that it is not important at all, that the manager's job is to shepherd employees though the documentation process and that the process is independent of the particular technology being written about. The manager provides estimates, ensures adequate staffing, tracks progress, and assesses how well each employee is performing. The problem with this approach is that it avoids understanding the business of the overall organization, what it makes and does, and how it achieves its overall mission. While you might be able to appreciate the mission without knowing the intricate technical details of the organization's products and services, you cannot completely comprehend how well it is being accomplished until you understand the means by which the organization is accomplishing it. Further, you will not be able to talk to the managers in scientific and technical groups in a way that shows that you acknowledge and support the products and the mission they seek to help fulfill.

Obviously, a technical communication manager has much to learn and to think about other than the finer points of the science/technology being covered by his/her employees. And in many environments it would take years of schooling to reach the same level of knowledge that the technical people have. Nonetheless, it is possible for a technical communication manager to learn the underlying principles by which the technology works and to understand how the products and services aid customers in achieving their goals. Only though this understanding can a manager truly learn how to provide the most value from the technical communication group to the overall organization. During my career, I have learned about environmental controls on smokestacks, explosive gas detection technology, diesel engine testing devices, power plant design and construction (coal, nuclear, and other methods), telecommunications, computer CPUs, database theory, and many other topics.

Several methods will help communication managers to learn at least the basics of the field in which they are working. First, take advantage of any internal training that is offered. In many scientific and technical environments, basic training courses are developed for training new technical employees about how the organization works on its products and services and about the underlying technologies used. Even if your organization teaches such courses at a level that may exceed your initial understanding, attend any classes you can. This shows everyone that you are eager to learn the technology, which immediately impresses them, and that you want to know how to contribute beyond merely creating documents. Second, ask your manager and other more experienced people for recommendations about books on the subject that are written for the lay person. Third, find out what magazines and technical journals the scientific/technical staff take, and subscribe yourself or read them in the organization's library, if there is one. In many industries, free journals are available if you simply fill out a data sheet about your responsibilities within the organization. Fourth, look for outside training and/or university courses that will give you the fundamentals. Finally, search on the Web for on-line training on the subject and for sites (sometimes of competitors) that will help you understand how the technology works.

Training Resources

There are many different ways to develop management skills. Given the limited time that most managers have for classroom training, they must seek other methods to enhance their knowledge and skill sets. The possible methods are similar to those discussed in Chapter 2, related to training for your employees. They include:

- Books
- Classroom training
- In-house training
- External training
- Distance learning
- University courses
- Computer-based training
- Magazines and journals
- Membership in professional organizations

In addition, there are other types of training available for management subjects, including:

- **Management Institutes**—Many organizations offer management training programs in various configurations. Some are offered via the Web, some via evening and/or Saturday courses, and some via intensive one- to four-week residency "experiences," wherein a group of managers receive all-day training at a specific site for several weeks.
- **University Management Programs**—Universities offer many of the same types of training experiences as listed above. In addition, many universities offer Management in Business Administration (MBA) courses and programs on campus, via the Web, and in special evening and weekend programs.
- **University/Corporate Partnerships**—Some universities and businesses set up partnerships with various configurations. Sometimes university professors come to the business to teach courses related to management. Some businesses contract with a professor or a university to modify an existing

course or design a new one specifically to fit their particular needs. Another model involves exchanges, wherein managers from a business come and teach university courses (or segments of them), and professors take a semester leave of absence and work on-site for the business.

For more information about all of the potential methods for obtaining management training, see the "Training" subsection in Chapter 2 and the list of resources in the back of the book.

Is Career Development Your Responsibility or Your Organization's?

As the preceding sections demonstrate, there is much that a technical communication manager must learn and must stay informed about. Should your organization provide all of the time and funding necessary to do so? When you become a manager, you must decide how much of your time (and money) you are willing to spend developing your knowledge and your skills. It is almost impossible to keep up with all of the areas described here during a 40- or 50-hour workweek. All employees should assume that they are at least partially responsible for their own growth and career development. Managers even more so. No matter how stable your current position seems, you must assume that some day you may work somewhere else. If you are not current with the managerial and technological developments in the technical communication field, you will have a much harder time finding a new position. A new manager must learn many new skill sets, but he/she must also learn a new way of thinking.

That thinking should include consideration of the larger, overall trends of the discipline as well as the shorter-term, more technical trends. Those larger trends point to dramatically changing roles for many jobs in general (Rifkin & Heilbroner 1995) and for technical communication jobs in particular (Johnson-Eilola 1996; Weiss 2002). Managers who simply keep doing things the way they always have are likely to find themselves irrelevant and expendable. The likelihood of working in the same job doing the same tasks for the same organization for a lifetime is very, very low. In the last 10 to 15 years technical communication has moved from being a profession that largely involved developing paper books to one that now largely involves developing on-line information. Many technical communicators do work that is far different from that of just a few years ago and, the odds are, in a few more years it will be far different from what it is today. Technical communication managers should consider the implications of such changes and anticipate how they will affect their functions within an organization and, in a larger sense, their careers. Managers should read general treatments of employment and the nature of work, such as Rifkin and Heilbroner's *The End of Work* (1995) and Reich's *The Future of Success: Working and Living in the New Economy* (2002) so that they can keep up with larger trends related to careers and employment.

Many employees refuse to spend one moment of their own time or one dollar of their own money on developing their knowledge and skill sets because they believe that their employers should pay for those things. This is a foolish mind-set, but one that, unfortunately, is common. For a manager, such a mind-set is a disaster. To an extent, being a manager is partly a job and partly a way of life. That does not mean that you spend all of your free time thinking about work-related matters, but it does mean that you spend some of your own time reading journals and magazines related to your field, that you spend some time attending professional organizations' meetings and functions, and that you constantly seek ways to develop and improve your skills. You cannot learn enough and improve your skills enough during the typical week at the office. If you fail to keep up with what is going on in management, technical communication, and your scientific/technological discipline, you risk becoming stagnant and not growing intellectually or professionally, and you risk having your skills atrophy to the extent that you are not employable.

Further, staying abreast of what is going on while constantly developing your abilities is very rewarding, whether you are able to apply your newfound knowledge immediately or not. While you will become a better manager and employee through constant growth efforts, you will become a happier and more complete person, too.

Use the Management Training Plan Worksheet to develop your own training plan for the coming year. Sometimes it is a good idea to develop such a plan over a two or three year period, so that you can see when and how to include all of the training you need. You should try to have some means for developing your skills for each of the five main categories, whether it be taking a training class, reading a book, subscribing to a journal, or other activity.

References

Anderson, P. V. 1994. Teaching technical communication managers about organizational management. *Publications management: Essays for professional communicators*. Edited by O. J. Allen and L. Deming, 213–228. Amityville, NY: Baywood.

Bush, D. W. and C. P. Campbell. 1995. *How to edit technical documents*. Phoenix: Oryx Press.

Dobrin, D. N. 1983. What's technical about technical writing? *New Essays in technical and scientific communication: Research, theory, practice*. Edited by P. V. Anderson, R. J. Brockmann and C. R. Miller. Farmingdale, NY: Baywood, 227–250.

Eisenberg, A. 1992. *Guide to technical editing: Discussion, dictionary, and exercises*. New York: Oxford University Press.

Finkler, S. A. 1996. *Finance & accounting for nonfinancial managers*. Paramus, NJ: Prentice Hall.

Hackos, J. T. 1989. "Documentation management: Why should we manage?" *STC 36th Annual Conference Proceedings*, Society for Technical Communication.

—— 1994. *Managing your documentation projects*. New York: John Wiley & Sons, Inc.

—— 1997. Using the information process-maturity model as a tool for strategic planning. *Technical Communication* 44(4): 369–381.

Hopwood, C. V. 1999. Staff training: Filling the gaps. *Intercom* 46(1): 18–21.

Technical Communication Management Worksheet

Technical Communication Management Training Plan

For each of the five main categories, list under "Development Method" what course, book, journal, etc., you will use to improve your skill and knowledge in that area for the current year. You might want to add more columns to the right to include future years.

General Skill	Specific Areas of Expertise	Development Method
General Management Skills	Project Management Time Management Personnel Management Financial Management Strategic Planning Problem Solving	
Tech Comm-Specific Management Skills	Document Management Information Architecture and Management Document Assessment Design of Document Libraries and Individual Documents Editing Estimating	
General Technical Skills	Computers/Networks/Communications On-line and Web-based Technologies Distance Learning and Delivery CD-ROM/DVD Production and Delivery	
TC-Specific Technical Skills	Publication Development Methods and Software On-line System Development Methods and Software html and Web-based Development and Delivery	

Industry- or Discipline-Specific Skills	Technology Involved in the General Industry
	    of the Overall Organization
	    (i. e., Telecommunications, Computer Software,
	    Pharmaceuticals, etc.)	

Houser, R. E. 1998. Create your personal training plan. *Intercom* 45(10): 4–9.

Johnson-Eilola. 1996. Relocating the value of work: technical communication in a post-industrial age. *Technical Communication Quarterly* 5(3): 245–270.

Reich, R. B. 2002. *The future of success: Working and living in the new economy.* New York: Random House Vintage.

Rifkin, J. and R. L. Heilbroner. 1995. *The end of work: The decline of the global labor force and the dawn of the post-market era.* New York: J. P. Tarcher.

Samson, D. C. 1993. *Editing technical writing.* New York: Oxford University Press.

Weiss, E. H. 2002. Egoless writing: Improving quality by replacing artistic impulse with engineering discipline. *Journal of Computer Documentation* 26(1): 3–10.

For Further Information

JoAnn Hackos, in "Documentation Management: Why Should We Manage?" (1989) and Paul V. Anderson in "Teaching Technical Communication Managers About Organizational Management" (1994) both stress the need for technical communicators to learn management concepts, pointing out that most technical communication training and education offers little to prepare communicators for management roles. Anderson also covers the basics of problem solving and moving toward decisions and action.

Hackos's article, "Using the Information Process-Maturity Model as a Tool for Strategic Planning" (1997) and her book, *Managing Your Documentation Projects* (1994), emphasize the importance of technical communication managers engaging in strategic planning rather than project-oriented, short-term focuses. The issue of STC's *Technical Communication* in which the Hackos article appears (Vol. 44, Issue 4, November, 1997) has several articles and case studies related to strategic planning.

David N. Dobrin's classic chapter, "What's Technical About Technical Writing?" contends that our discipline is indeed technical and that communicators should learn something about the technologies for which they create documents.

Catherine V. Hopwood in "Staff Training: Filling the Gaps" (1999) explains methods for developing effective training systems for technical communicators. Rob Houser's article, "Create Your Personal Training Plan" (1998) describes how technical communicators can develop specific plans for their professional growth.

Questions for Discussion

1. Should an employee's organization be solely responsible for providing the time and the funds for all of that employee's professional growth? Should all training and learning happen on company time and at company expense?

2. To what extent is being a manager a job and to what extent is it a way of life? If it is only a job, then training for it and thinking about it should be done on the job. If it is a way of life, then training for it and thinking about it go on continuously, on the job and off.

3. What is more important for a new manager? To keep current with the technologies being used by his/her employees, or to develop better general management skills?

4. Which of the five main types of training require ongoing, continuous learning and which lend themselves to one-time solutions like training courses or books?

5. Is it important for technical communicators to know something about the science and technology about which they are writing? Can they do a better job of user advocacy if they do not know too much? On the other hand, should they become experts in the science or technology, even learn as much about it as the subject matter experts? Should technical communication managers be experts?

6. Should a technical communication manager know more about the tools used by his/her group than the members of the group? How does a manager stay ahead of everyone else if he/she is no longer involved in developing documents?

CASE 5

Management and Technical Skills Development Plan

The Management Situation

Because your manager at Aardvark is not a technical communicator, you have been asked to prepare your own program for developing your skills and knowledge. Your manager wants you to improve your management acumen, but also wants you to keep current with changes in your discipline.

You have been asked to prepare your own Management and Technical Development Plan and submit it to your manager for approval. The plan should be for the next three years and can include up to three training courses (in-house or external) per year.

You will have to weigh several variables as you design your plan. First, as a new manager, you need to spend as much time as possible learning about the job by doing it. So, you cannot afford to take too much time off for training. Second, you will have to balance among five main types of training that you need to acquire:

- General Management Skills (Time Management, Budgeting, Project Management, Personnel Management)

- Technical-Communication-Specific Management skills (Technical Communication Management, Documentation Storage and Retrieval Management, Production Management, etc.).

- General technical skills (Trends in technical communication, On-line and Web-Based Technologies, Distance Learning and Delivery, CD-ROM Production and Delivery, etc.).

- Technical-Communication-Specific Technical Skills (Framemaker for document development, RoboHelp for on-line help development, html, tools for developing Web pages and sites, etc.).

- Industry- or Discipline-Specific Technology (knowledge of the science and/or technology of the specific industry in which you are working; for example, communications, computers, pharmacology, etc.).

Another variable is the method you will use to cover each of the areas above. There are many ways to develop skills other than through in-class training sessions, including:

- Books
- Distance learning via the Web
- Self-paced training using computer-based tutorials
- Magazine and journal subscriptions
- Membership in professional organizations and attendance at conferences

Because you cannot cover the five main areas you need to cover by attending training courses alone, you will have to use some of the alternative methods to improve your skills in at least one or two of those five areas. You will also have to consider which of the areas to place higher priority on. What is more important for a new manager? To keep current with the technologies being used by his/her employees? To learn more about management theories and practices? Your answers to these questions will determine how you cover each of the five main areas.

The Assignment

Using the Technical Communication Management Worksheet supplied in this chapter, prepare a training plan for the next three years. You may have to reformat the sheet and use several pages to accommodate all of the methods you decide to use. Also, your manager will want to know the cost of the methods you propose to use.

Helpful Hints

Try some of the following sources for more information about management and technical training.

The American Management Association provides classroom and self-study courses in numerous management areas. See http://www.amanet.org/

The advertisements from training and consulting companies in the Society for Technical Communication journal, *Technical Communication.*

Your organization's internal training courses.

Courses available at nearby colleges and universities.

Franklin/Covey is another company that offers a wide range of management training materials. See http://www.franklincovey.com/.

See the training organizations listed in Chapter 9 for companies that offer training specifically in technical communication.

Evaluation Criteria for Case 5

Have you covered each of the five areas with adequate training and development methods?

Have you justified the decisions you made with sufficient explanations and sound reasoning?

If you have weighed one or two of the areas more heavily than the others, have you explained why you did so?

Have you found training and development resources that will do an adequate job of keeping you informed?

Have you dealt with the dilemma about company time versus your own time?

Managing Yourself

Mentoring

Management employees are wise to find and develop mentoring relationships with more experienced managers. It is simply impossible to go through training courses and management books fast enough to learn everything that one should know about management. Further, even after doing so, managers will not learn about the political and social peculiarities within their own organizations. The management books may all say to do something that is frowned upon in a specific company. For those reasons, it is a good idea for a manager to seek mentoring from managers within the internal organization. The mentor does not necessarily have to be from a technical communication background to help a less-experienced manager with the specific politics and management methods of the organization. Some organizations even establish formal mentoring systems wherein the mentors must be from other divisions or departments, so that their management staff learns how other groups work, and to avoid developing insular outlooks within any single department.

What does a technical communication manager do about technical-communication-related mentoring when there are no other managers within the organization who have communication backgrounds? There are two main sources of such specific mentoring. One is the STC Management Special Interest Group (SIG) mailing list, at http://www.stcsig.org/mgt/index.htm. This mailing list is a boon to technical communication managers who cannot get expert advice from within their own organizations. The list is monitored by dozens of experienced communication managers who happily answer questions and provide advice to less experienced managers who pose questions. Another way to get mentoring is to join one's near-est STC chapter and get to know members who are in technical communication management positions at other organizations. Some of them may be willing to act as unofficial mentors, answering questions and helping with difficult problems. Being a lone technical communication manager is a difficult job, one that should not be attempted without the help of others. Sometimes simply knowing that others have experienced the same kinds of problems one is dealing with offers considerable solace. Use of the mailing list and meeting with or e-mailing other managers can help the isolated technical communication manager work through difficulties and learn new job skills.

Time Management

Time management is a critical skill for any manager, but especially for technical communication managers. Nearly all technical communication work is time sensitive; it involves deadlines. Missing deadlines is the cardinal sin for a technical communicator, as doing so causes lost revenues, lost esteem, and lost credibility (and, occasionally, lost jobs). For some reason, people remember for a long time when you have missed a deadline, so it takes years and many on-time deliveries to compensate for even one deadline failure. Because many communication groups work on multiple projects at once, a technical communication manager may have to track dozens of milestone and deliverable dates simultaneously. Further, the multiple projects may mean communicating with dozens of people, resulting in e-mails, telephone calls, voice mail messages, pager beeps, meeting agendas and minutes, and other incoming requests for your time and attention.

Without the organized, persistent use of a structured time management system, a manager can get in trouble in a hurry. A good time management system should allow you to keep up with your personal action item list (or "to do" list), your schedule (daily, weekly, monthly, annually), and a directory of contacts with phone numbers, e-mail, and any other contact information you need. The system should allow some method for imposing priorities on your "to do" items so that you do not simply have one long list with mixed short-term, long-term, and life-time goals. It should also give you a way to establish priorities so that you know what you will work on first each day.

There are three primary methods by which most people maintain time management systems: paper-based, computer-based, and PDA-based. Before deciding which is the best one for you, it is a good idea to ask other managers what systems they use, especially if those managers are people who seem to manage their time well. While the various types of systems have their individual advantages and disadvantages, you should follow the system that other people in your organization use, especially if it is a standard and is supplied by the organization.

A paper-based system has the advantage of being portable, perhaps even fitting in a pocket; of allowing you to insert reminders, notes, papers, and business cards into it; and of being usable in almost any environment. It has the distinct disadvantage that it does not automatically update, so that you have to rewrite your "to do" list every day. It also cannot handle repeating schedule events. Rather, you must write each one in separately. It also does not automatically remind you of appointments, requiring you to consult it regularly.

A computer-based system has the advantage of automatically updating itself (you don't have to re-copy all of those action items every day), of holding huge quantities of information without becoming too heavy to carry, of popping up windows and or beeping to alert you to important schedule events, and of automating tasks like addressing e-mails to a group. A major disadvantage is the lack of portability, unless you have the system on a laptop, which is still not as convenient as a small notebook or a PDA. You must input everything, so that the usual set of notes, stickers, and business cards that gather in a paper version must be entered as you get each one.

A PDA-based system is also portable and convenient, is capable of holding a considerable quantity of information, can by synchronized with a computer-based system, and, potentially, can communicate from anywhere. Its major disadvantage is that inputting information is awkward and slow. While you can also buy a lightweight keyboard to expedite entering data, that reduces the portability and hence much of the PDA's advantage.

Many people opt for combinations of systems. Because PDA files can by synchronized with the files on a desktop or laptop computer, many people keep their time management systems on their main computer and download it to the PDA when they must be out of the office. Others prefer to print out a daily list of appointments and "to do" items to put in a notebook.

Fortunately, there are many products available that help make establishing a time management system much easier. The two most frequently used time management systems are those offered by Day-Timers (http://www.daytimer.com) and Franklin/Covey (http://www.franklincovey.com). Day-Timers offers paper time management systems ranging in size from pocket editions up to 8 $1/2$ x 11-inch binders. They also provide software versions for use on a PC or a PDA, including the ability to print each day's lists for inclusion in one of their binders, potentially providing the advantages of all three types of systems in one unit. Likewise, Franklin/Covey offers both paper- and computer-based solutions, including software designed to run on a wide array of machines and to coordinate with other software, such as Microsoft's Outlook. The Franklin/Covey system operates on the principle-based system espoused by Stephen Covey in his books *The 7 Habits of Highly Effective People* (1989) and *Principle-Centered Leadership* (1991) and Smith and Blanchard's *What Matters Most* (2000). The Day-Timer series is oriented more toward classic time management tracking of actions items, contacts, and scheduling.

Both systems offer numerous versions of their time management systems and the books and training upon which they are based. It is a good idea to get some training and/or to read books about time management. Using one of the paper or software solutions to track appointments and contacts will not necessarily help you to manage your time more effectively, unless you understand how to use the system effectively.

Time Inventory

One of the methods most of the time management systems teach is the time inventory. Here, you inventory how you are spending your time each day for a typical workweek. You do this for one week, assuming that it is a reasonably typical workweek for you. Try to pick a week for the inventory when you do not have any extraordinary events, such as all-day, off-site meetings, planned. Doing the inventory entails preparing a chart and tracking how you spend every minute of the day. The goal is to get an idea of what percentage of your time you spend at various activities during the week, and which parts of the day you normally devote to the various duties. While it does not have to be accurate down to the last minute, the inventory should allow you to see where you spend your time. You should honestly include time spent on coffee breaks, bull sessions, and non-work-related activities. You should also include all overtime you put in, including phone calls, on-line activities, or reading you do away from your office during evenings and weekends. You are trying to discover how you spend your working hours, how you might spend them more effectively, and which activities you might be able to eliminate or curtail. As a technical communicator who works on intellectual property in a pursuit that is highly creative and challenging, you want to eliminate as many of your rote, repeated, physical activities as possible to allow yourself to focus more on planning, tracking, and completing your projects. Doing a time inventory is a first step in working toward establishing a good time management system and toward improving your work life. Use the Time Inventory Worksheet to record your activities for a week, following these steps:

1. Start on Monday morning and record every time you change activities. This means that you will carry the inventory with you throughout the day and will briefly interrupt activities to record what you're doing. Yes, it is a distraction, but you will only be doing it for a week. If you wait until the end of the day, you will forget five- and ten-minute segments that occurred earlier in the day, even though you may think that you won't.

2. Record the time in as much detail as possible. If you spend 30 minutes on e-mail, do not simply put down "e-mail, 30 minutes" but calculate how much of the time was on essentials and how much was spent reading and deleting junk e-mails from inside and outside your organization. If you spend two hours in a meeting, calculate how much of the time was of value to you and how much was wasted.

3. Try to spend the week doing what you normally do, even though keeping the inventory will make you start seeing wasteful activities while you are in the process. The inventory will be more realistic if you continue doing what you would normally do.

4. At the end of the week, analyze the inventory. You can break it down in several ways:

    How much time did you spend on your top priority for the week?

    How much time did you spend on your long-term goals and objectives?

How much time did you spend on communication via e-mail? Telephone? Meetings? Personal contacts? How important were those contacts toward achieving your goals?

What was the longest period of uninterrupted time you had all week? This may come as a shock.

When did you get the most accomplished?

When did you get the least accomplished?

Is there someone who is taking up an inordinate amount of your time?

Did you complete lots of small action items but make no progress toward the larger ones?

Who is interrupting you and how often? Are those interruptions necessary?

How much time did you spend searching for information, whether it be paper or on-line? Is there a better way to store it so that you can reduce this time?

Make a list of your top priorities and calculate what percentage of each you completed during the week.

See if you can classify the time periods into larger groups, such as communication, personnel issues, meetings, finances, etc. This can sometimes help show you important trends in what takes up your time.

There are many other questions you can ask as you analyze the results. You are not trying to drain every last second of non-work-related conversation out of your workday, but you are trying to see what activities and people take up your time, and if you can reduce the time spent on those that are not critical and spend more time on the important issues.

You are trying to avoid what Mackenzie (1997) calls the "time trap" and others have labeled the "activity trap," wherein you work hard and complete many minor tasks while never taking the time for planning or completing the larger, more important tasks. Because you never do any planning and prioritizing, you stay stuck in the trap of furious activity without accomplishing the larger goals you would like to.

Biological Clock

Whether you believe in the concept of a biological clock or not, the fact is that most of us have times of the day when we work more effectively than others. One way to improve your productivity and the satisfaction you get from work involves analyzing which times of the day you do your best and fastest work. Doing the time inventory in the previous section can help, but you need to analyze the results carefully. Upon arriving at work, most of us first read our e-mail, listen for any voice mail messages, and look over any new paper mail. However, if you are one of those people who work most effectively early in the morning, you might want to consider bypassing such mundane activities so you can focus on the more difficult and complex tasks you need to complete. Hence, just because the time

Technical Communication Management Worksheet

Management Time Inventory

"To Do" Items for Today:

Task/Goal	Rank	Deadline Time

Time Spent Today

Time	Activity	Minutes	Priority	Comments

inventory shows that you spend several hours a week handling communication tasks first thing in the morning, it does not mean that that is the best time for you to do such tasks.

You might want to consider which times of the day you routinely do your best work and which times you seem to work more slowly. You can save the more mundane tasks for those times, freeing up your better times of the day for more important tasks. Because technical communication is at least partly a creative activity, you should strive to complete the more creative tasks, such as analyzing and planning, during times of the day when you are most effective at them.

Be Here Now

A serious result of the time trap is that nothing you work on ever gets your full attention and your best effort. Even those things you do complete are done hastily and haphazardly because, with all of the interruptions and changes in priorities during your workday, you can never give your full attention to anything. So, you struggle with the tension of knowing that you are doing mediocre work but that you can't do anything about it. This is a devolving process that eventually leads to burnout. One way to start climbing out of this hole is to adopt the policy of paying total attention to whatever you are doing. This means that there may be times when you put a "Do Not Disturb" sign on your door, close it, turn off the telephone, and bore in on what you need to work on. It also means that when you go to meetings you turn off the cell phone and the pager, and that you concentrate totally on what is happening in the meeting. As I have said elsewhere (see Chapter 2), personnel meetings with your employees should not be interrupted, no matter how important that phone call is that you are expecting.

While you cannot and should not completely eliminate interruptions, there are times when you need to reduce them. O'Conaill and Frohlich (1995) report that in over 40% of the cases when employees are interrupted they do not return to the task they were originally working on.

This is sometimes called the "be here now" principle. That is, you devote full attention to doing whatever you are working on, without allowing constant interruptions to interfere. To an extent, the intrusive communications technology available to us makes "being here now" difficult, as we are constantly interrupted by ringing office telephones, vibrating cell phones, beeping pagers, chiming e-mail boxes, etc. As your time inventory may show you, to spend more than a few uninterrupted minutes on any task, you may have to disable some or all of the technology, or simply ignore it. If you are one of those people who are constitutionally incapable of letting a telephone ring without answering it, leave your office and find a place where you can concentrate without the interruption of the telephone. Consider turning off your e-mail program's chime and limiting your e-mail sessions to one at the beginning of the day, one at lunch, and one at the end of the day.

Delegation

One of the common causes for getting caught in the time trap is inexperience and/or reticence in delegating responsibilities. Many managers, especially newer ones, fear delegation because they believe that they should be able to do everything themselves, or they believe that only they can do the jobs properly. If you are thinking in either of those ways, you need to make some changes to the way you are managing your group. If your group is taking on enough work, you should certainly not be able to do it all yourself. Further, if you want your employees to grow professionally and to learn how to handle greater responsibilities, you have to assign those responsibilities to them, offer whatever help they need, then back off and let them get the jobs done. This is difficult when you first do it, but you learn quickly which employees have the capacity to get things done and which do not. Also, you want people in your group who can do some of the work that you have to do. You should want people reporting to you to want your job and who are capable of doing it, or at least large pieces of it. The way to get such employees is to delegate responsibilities to them. It is important to give an employee to whom you are delegating a task all of the details you possibly can about what the task requires and what you expect the outcome to be. While some employees may be very good at discerning what results you want, many will not be as good at it. The more explicitly you can describe the desired outcome, the more likely you are to get that outcome.

When you are going to be away from the office for several days, whether on business or vacation, you should delegate your responsibility to someone in your group. This can be a very effective way of training someone in how to do your job and testing whether that person might be ready to move into a supervisory position. Delegation policies vary among organizations, but if yours allows or requires delegation, give careful thought to who you want to be your delegate and how you can prepare that person for the role. You should describe for your delegate any crises that are ongoing, any important deadlines or deliverables that will occur during the period, and anything beyond the normal routine that you expect to happen. You should also encourage the delegate not to engage in any serious personnel matters, but to defer them or, if action is required, to get your manager involved.

Setting Objectives and Goals

As discussed in Chapter 2, every employee, including every manager, should have a set of goals and objectives. In a well-managed organization, these goals and objectives will be tied to the department's and the organization's overall goals and objectives. For a technical communication manager, it is especially important to tie personal goals and objectives to the larger aims of the department and the overall organization. Because communication managers' roles are often very different from everyone else's, and because other employees often assume that we do not understand larger goals, it is important to tie our goals to theirs and to demonstrate how achieving our goals helps the overall organization to achieve its goals.

If you report to someone who is not from the technical communication field, you will want to work with that person to develop your objectives, lest you receive a set that does not realistically reflect what you hope to accomplish. If your organization does not require that you set personal objectives, it is a good idea to do so anyway. Even if they go unobserved by your manager, you will have the satisfaction of knowing which ones you achieved, and you will also have the guidance that a set of objectives provides about how you assign priorities to the tasks you face every day. A good set of annual objectives for a technical communication manager should include:

- An objective related to ontime delivery
- An objective related to improving document quality
- Several other objectives that relate to your organization's specific work and that are challenging but achievable
- An objective aimed at improving the processes or efficiencies in the way your group works (such as writing a style guide or a set of templates)
- At least one stretch objective that will be difficult for you to achieve but that will demonstrably either directly meet one of your organization's goals or improve your skills as a manager

Annual objectives should have specific dates assigned to them, so that you don't have until December 31 to complete them all. They should also include the means by which you will measure whether you have successfully completed the objective.

If your manager does not assign objectives and does not hold interim and final performance reviews each year, then hold them with yourself. Do a mid-year review and see how many objectives you have completed and what your prospects are for completing the remaining objectives during the year. At the end of the year, review how well you have done on the entire set of objectives. This is often a time when you get some perspective about time management problems because you may realize that you have engaged in "fire drills" all year but have not accomplished as many of the larger, more important goals as you had hoped. Objectives are discussed in more detail in Chapters 1 and 2.

Managing Your Life

Here is an interesting experiment. On a blank sheet of paper, define yourself in 25 words or less. Do not read the following paragraph until you have done so.

After your first self definition, take out another sheet and define yourself again. This time do not use any terms associated with work (job titles, roles, assignments, etc.) or with family roles (mother, father, uncle, etc.).

What results did you get? Some people have a very difficult time completing the second type of definition because they may have allowed their self definition to become almost completely wrapped up with their professional lives. If you had trouble coming up with 25 terms about yourself without referring to your work, you should do some of the thinking discussed in Chapter 1. People who are better rounded and who have progressed closer to the top of Maslow's hierarchy will have two definitions that are similar. They may have had to drop a couple of terms from the first definition when they wrote the second one, but their definitions will be similar. That is because they define themselves and think about themselves in ways that extend beyond work and family matters. Most of them have hobbies and interests that provide them with creative, intellectual, and/or physical outlets. For technical communication managers who are often working on long-term projects, such outlets may be necessary. Further, technical communication managers must often deal with the difference between being directly involved in creating documents when they were employees and being only indirectly involved as managers, with the result that their sense of satisfaction and reward is not nearly as great. Hobbies, college courses, participating in civic groups, coaching soccer teams, and other similar activities can provide a sense of accomplishment that many managers miss when they stop being directly involved in producing documents.

Paradoxically, many managers are more productive and effective when they work fewer hours and devote more of their time to outside activities. As happier, more rounded people, they become better managers whose thinking and decision making are clearer and whose ability to help others set and pursue goals is better. Mackenzie (1997) has an excellent extended discussion about workaholics, including ways to measure whether you have become one. As he points out, some people become workaholics because they have no choice, they may have to work two or three jobs for 60–80 hours a week to pay the bills. Others work constant overtime because they are inefficient and can accomplish 40 hours of work only by spending 60 hours doing it. And others are truly addicted to their jobs, which they may even enjoy.

While it is beyond the scope of this book to treat the subjects of workaholics and burnout, if you think you are devolving in that direction, read Mackenzie (1997) and others on the subject and seek some counseling through your organization's employee assistance department or through an external counselor.

Most of us expect to have sets of objectives at our jobs. In fact, we might be offended it we were not given a specific set to work toward each year. Ironically, though, many people have no objectives at all for their private lives. Most time management systems, for example, Covey (1989), suggest that you begin any such system by recording your own personal objectives prior to recording those that are required of you at work. Doing so can help give a perspective on what you hope to accomplish both personally and professionally, which can help you make decisions about how much time and effort to devote to personal life and to work life. Using Covey's system provides the further bonus of using principle-based management methods to balance one's life at work and at home.

References

Blair, G. M. 2002. Personal time management for busy managers. http://www.see. ed.ac.uk/~gerard/Management/art2.html. (November 17, 2002).

Covey, S. R. 1989. *The 7 habits of highly effective people*. New York: Simon & Schuster.

Covey, S. R. 1990. *Principle-centered leadership*. New York: Fireside.

Mackenzie, A. 1997. *The time trap*. New York: Amacom.

O'Conaill, B. and D. Frohlich. 1995. "Timespace in the workplace: Dealing with interruptions." In *Conference Proceedings: 13th Annual Conference on Human Factors in Computing Systems*. Denver, CO: Association for Computing Machinery (ACM) Special Interest Group in Computer-Human Interaction (SIGCHI).

Shea, G. F. 1999. *Making the most of being mentored*. Los Altos, CA: Crisp Publication.

Smith, H. W. and K. Blanchard. 2000. *What matters most: The power of living your values*. New York: Simon & Schuster.

For Further Information

Stephen R. Covey's books, *The 7 Habits of Highly Effective People* (1989) and *Principle-Centered Leadership* (1990), treat managing oneself related both to work and to private life.

The American Management Association offers an on-line course led by Peter F. Drucker titled *Managing Oneself: Taking Responsibility and Developing Opportunities,* at http://ama.corpedia.com/welcome/product.asp?product=8101.

Gerard M. Blair provides a good introduction to time management in his article at http://www.see.ed.ac.uk/~gerard/Management/art2.html .

The AMA also offers a seminar in time management. Other seminars in time management can be found on the Franklin/Covey Web site at http://www.franklincovey.com and the Day-Timers site at http://www.daytimer.com. In addition, these organizations market full lines of time management software, books, folders, and other material. Mackenzie in *The Time Trap* (1997) explains concepts related to time management and provides numerous methods for managing time more effectively.

The Mindtools Web site on time management provides information about activity logs and other time management issues, at http://www.mindtools.com/pages/article/new.

Nearly all of the literature written on mentoring is written from the point of view of the mentor or the manager who is setting up a mentoring program. Gordon F. Shea treats the subject in *Making the Most of Being Mentored: How to Grow from a Mentoring Partnership* (1999).

Questions for Discussion

1. Do you currently use a time management system? What kind? Does it include goals for the long-term, intermediate-term, and short-term, or only those immediate tasks that you need to accomplish in the next few days?
2. Are there certain times of the day when you are more productive and when your mind seems to be at its sharpest? Does this happen naturally or does it require external stimulants, such as caffeine? What kinds of activities do you engage in during those periods?
3. Using a sheet of paper, record for one day all of the times that you are interrupted by telephones, e-mail chimes, people walking in, etc. How many interruptions did you experience? Divide the number of minutes you monitored by the number of interruptions to find out how many minutes you can expect to devote to a task without interruption. Is that number higher or lower than you expected?
4. Do you have a set of written objectives for your personal life? For your work life?

CASE 6

Managing Your Time

The Management Situation

You have decided to get more control over how you spend your time. Whether you are employed or not, inventorying how you spend your time can be a valuable exercise. This case, then, deviates from the others in the book in that it has you do your own time management inventory rather than working on a simulated management situation. Your goal here is to perform a time inventory, to analyze it, and to write recommendations based on it.

The Assignment

Re-read the section on time management. Using the Time Inventory Worksheet supplied in this chapter, or a similar one that you create, record all of your activities for one week. If you are employed, record the times that you are at work and the times that you spend on work related activities at home or at other places. If you are not employed, record all of your activities throughout the entire day. Remember, you will get a much more accurate picture if you keep your inventory sheet with you and record activities as you complete them, rather than relying on your memory to re-create them at the end of the day.

At the end of the week perform an analysis on the results, grouping similar time periods together (e.g., meetings, communications, leisure, reading, writing, watching TV, etc.). To do this analysis, you need to be as dispassionate as possible, to step back and look objectively at the results of your study. Assume that you are looking at the results of a friend's use of time rather than your own. After analyzing the data, write a set of recommendations about how the person you have just studied could better spend his/her time. Note that there are no right or wrong results here. Some people will recommend less time on certain types of activity, while others will recommend more time for the same activity.

Your report should have four sections: Introduction, Time Inventory, Analysis, and Recommendations. Submit the report along with legible (to the extent possible) copies of all of your time inventory sheets for the week.

Helpful Hints

Use the Time Inventory Worksheet provided in this chapter.

See the Mind Tools Web page on time management for more about activity logs and other time management issues, at http://www.mindtools.com/pages/article/newHTE_03.htm.

For a good introduction to time management, see Gerard M. Blair's article at http://www.ee.ed.ac.uk/~gerard/Management/art2.html.

For time management information aimed specifically at university students, see York University's http://www.yorku.ca/cdc/lsp/tm/time.htm.

For complete time management systems, including books and training courses, see http://www.daytimer.com and http://www.franklincovey.com.

Evaluation Criteria for Case 6

Does your inventory account for time in enough detail so that you can analyze how you are spending it?

Have you classified like kinds of activities into logical groupings that help you analyze how you spend your time?

Does your report thoroughly analyze how you have spent your time? Does the analysis show you which of your top priorities/goals you completed and where you got sidetracked from doing so?

Do your recommendations follow logically from the inventory, classification, and analysis?

Managing Your Boss

Introduction

Having a positive, healthy relationship with your manager can make the most mundane of jobs seem rewarding. And having a poor relationship can make any job seem like a prison sentence. As Marcus Buckingham (2001) of the Gallup Organization puts it, "Our research tells us that the single most important determinant of individual performance is a person's relationship with his or her immediate manager. It just doesn't matter much if you work for one of the '100 Best Companies,' the world's most respected brand, or the ultimate employee-focused organization. Without a robust relationship with a manager who sets clear expectations, knows you, trusts you, and invests in you, you're less likely to stay and perform."

Many books and articles in the general management press explain various methods for working well with your boss. In the case of technical communication managers, there is often an extra twist involved because many of our immediate managers are people who have never been technical communicators and who have other backgrounds and interests. They may be overseeing our operation against their will, which, obviously, presents special problems to the technical communication manager (Taylor 1989).

This chapter presents some general ideas for working well with your boss. It also describes some of the special issues concerned with technical communicators who report to people from other areas.

Empathize with Your Boss's Situation

One of the first steps in getting along well with someone is to be able to empathize with that person. To do that, you may need to do some self-examination before you try to examine your boss. In particular, you want to consider your own set of assumptions about people in authority. Did you get along with both of your parents? Teachers? Others in authority? Do you assume that all managers have sold out and prostitute themselves for the almighty dollar? What are your preconceptions about authority and management? Are they influenced by previous experiences that might not apply to your current position?

After you analyze your general feelings about authority and management, you should then consider whether you have any preconceived notions about the kind of person that your particular boss is. Do you not like short people? Tall people? Bald people? Graduates of State U.? Men? Women? This is a time to force yourself to confront your own biases and prejudices because they may be partly responsible for poisoning your relationship with your manager. Granted, many managers deserve their fair share of loathing and contempt. However, if you have a preconceived set of ideas about your boss, you need to examine why, to try to exorcise yourself of them (including, perhaps, some of the things you were taught as a child), and to give your boss (and yourself) a chance for a successful relationship.

Once you analyze your own mind-set regarding your boss, you can begin to analyze his/her situation. Ask yourself a few fundamental questions:

Who is his/her boss? What is it like to report to that person? How much positive reinforcement do you think your boss gets? What do you think he/she gets rewarded for? What kinds of pressure are applied from his/her boss and how are they passed on to you? Is your boss's boss highly regarded? Sometimes considering your boss's boss can give you a new understanding about some of your own boss's behaviors, mood swings, and priorities.

What constraints and requirements does he/she work under? Is your boss free to fix problems that need to be fixed? Does he/she have the power to affect policies, especially those that affect your operations? Does he/she have requirements that have higher priority than those of your technical communication group?

What priorities and results are important? What tasks is your boss charged with accomplishing? What results make him/her happy and which ones cause consternation?

How is performance measured? What do you think will get your boss an outstanding performance evaluation? A poor one? Is your boss's performance based on financial performance? On-time deliveries? Cutting costs? Increasing revenues? How do you think the technical communication function contributes to or detracts from the criteria against which he/she will be measured?

Many people have such an aversion to authority figures or to all management personnel that they find it very difficult to empathize with someone from that "enemy" camp. Having that mind-set condemns them to careers that are almost guaranteed to be miserable because they are never going to have a positive relationship with their manager, even if they try to fake it. It is a self-defeating, downward spiral to dislike any boss, blame him/her for why you are miserable at work, react by disliking him/her more, and so on. Even if you dislike your boss for legitimate reasons, trying to empathize with his/her situation can help you have as positive a relationship as possible.

Get Your Own Job Done First

The best way to have a good relationship with your boss is to get your job done and done well. That makes the boss's life easier and it also expands the amount of influence you have with him/her. Technical communication managers should be aware that the circle of responsibility a manager assumes affects the circle of influence that he/she gains. You have probably noticed that some managers seem to have wide spheres of influence, working with many other groups and having an effect on the policies and thinking of many in the organization. Others, even at the same management level, seem to have little or no influence at all. They simply and repetitiously take care of their niche, but they do not seek anything outside of it. This is not a case for empire building. Rather, it means that good managers look for ways to improve all aspects of the larger organizations in which they work. When they do so, they gain greater credibility and more influence over what goes on in the organization. As Bryan (1994), Hackos (1994), and Plung (1994) have pointed out, the mature technical communication group takes care of its own business by turning out quality documents on time, but it then works with other groups to improve the organization's overall results. For example, it may help with proposals, marketing materials, usability testing, Web site design, information management, and in many other ways not directly described in the technical communicator's original group charge to write documents. The more influence you have in the organization in general, the more influence you will have with your boss. Further, your boss will see you as someone who looks for and solves problems rather than someone who merely complains about them or who is a drain in some other way.

Some people, whether consciously or subconsciously, do not do their best work when they do not like their bosses. Some go so far as to sabotage their own work, thinking that it will reflect poorly on the boss and perhaps contribute to his/her departure. What it usually does instead is to reflect poorly on the employee and not the manager. Bosses got to be bosses because they understand how politics work, so they are likely to know how to make poor work reflect on others and not on themselves. A technical communication manager should never let the group's work be compromised because he/she doesn't like the boss.

If you become a reliable and trusted employee who gets things done well and on time, your circle of influence with your boss will increase. Your boss will listen to you and, eventually, will seek advice and suggestions from you.

Develop Credibility

It is important for a technical communication group to develop credibility, especially in scientific and technological environments where other skills may be more generally valued. You develop credibility with your boss by reliably completing your assignments on time, within budget, and with the highest quality possible given those constraints. You develop credibility with your boss by helping him/her accomplish the goals of his/her department. If there are no general goals associated with communication, then you should find ways to help with the other goals (and you should see to it that there are some explicit goals for technical communication).

You also develop credibility by not engaging in what Covey (1990) calls the "metastasizing cancers" of the workplace: complaining, criticizing, comparing, and competing. If every communication you have with your boss involves complaining about developers who get behind schedule, criticizing SMEs who don't get review copies back on time, or complaining about how poorly you are paid compared to the "techies," your boss will come to see you as a negative influence, as an energy draining individual who leaves him/her feeling worse than when you got there. Like it or not, many bosses view those who report to them as either pluses or minuses, as people who consistently make positive contributions that lighten the load or as people who constantly make the load seem heavier and more burdensome. Even if your boss begins to engage in one of Covey's four C's, try to steer the conversation somewhere else or to get out of it as soon as possible. You do not develop credibility with your boss by complaining all of the time. You develop it by bringing up problems, suggesting solutions, and then implementing them.

Likewise, avoid the us-them mentality that affects almost all groups (Bryan 1994). Even if your boss freely engages in complaining about other internal groups, try to avoid joining in. It is very easy for a technical communication manager and group to engage in the us-them mentality because communication groups often have very different backgrounds and interests than others in a scientific or technological environment. If your boss is not a technical communication person, this us-them mentality is going to seem especially unprofessional, immature, and foolish, and is likely to ensure that the technical communication group stays marginalized. Engaging in us-them thinking shows your boss that you do not understand or value the goals of the overall organization, which is always a serious mistake for a technical communication manager. Instead, you want to show an understanding and appreciation for the larger issues rather than for only your own concerns. You want to show that you understand the overall organization goals and to seek ways to work cooperatively with other groups within the organization, even when they make it difficult to do so. It is fine to seek your boss's assistance in that endeavor, but it is never a good idea to complain to your boss about other groups.

Become a Problem Solver Rather Than a Complainer

Bosses indeed tend to think of employees as either positive contributors who solve problems and contribute to the organization's objectives, or as negative contributors who complain about problems without ever solving any and whose day-to-day interaction saps time and energy without adding much.

If you need to present a problem to your boss, you should also present solutions and their advantages and disadvantages. (See Chapter 4 on communicating with your immediate manager.) This is especially important for a technical communication manager who reports to someone who did not come from the technical communication world. In that case, your boss may have trouble understanding why the problem is a problem. You will need to quantify how it affects results, in dollars or in days added to the schedule, rather than simply complaining that it is an inconvenience. You will then need to suggest the possible solutions because your boss may simply not know what the best ones would be. It is important that a technical communication manager help a non- technical communicator boss to solve problems rather than complain about them and then push the boss to solve them on his/her own. You want to collaborate with your boss in fixing things rather than demand that your boss do so. You also want your boss to be well informed about communication problems and to understand the implications and costs associated with not fixing them.

Present New Ideas Gradually and Prove Their Value

Many employees demonstrate a naïve belief that if an idea seems good to them it ought to seem good to their bosses also. Further, they seem to believe that they should be able to explain the idea to their boss in five minutes and to receive his/her undying gratitude, admiration, and approval. There are two things wrong with this approach. First, the person with the new idea often underestimates how long it took to arrive at it, assuming that the moment of revelation occurred easily because the answer was so obvious. In reality, the solution might have occurred after many hours or even weeks of rolling the idea around and considering many possibilities before the solution finally emerged. Second, the boss gets paid to have a different perspective and to view ideas with a different set of criteria than do the people who report to him/her.

This is particularly a problem for technical communication managers who report to someone not from the communication field. While a solution may seem crystal clear from a technical communication perspective, it may not be nearly so clear from the other perspectives with which the boss views it, especially if it is going to incur costs that are not offset in some way. Therefore, it is important that technical communication managers present new ideas to the management above them as proposals that include the costs of not doing anything, the costs and payback of each of the possible solutions, and recommendations for a particular solution and an implementation plan for putting it in place. Before proposing to change processes or products, a technical communication manager will need to use excellent communication skills in analyzing a management audience and its assumptions and priorities.

It is often better to present new ideas slowly, to set the stage for them first by giving the boss a background report or finding an article or two that speaks to the issue. If it took you awhile to arrive at conclusions about an issue, it will take the boss awhile too, so you need to give him/her time to digest new information about the subject and to develop some of the same background understanding you had before you arrived at your conclusions. Above all, second-level and higher managers like to see data and proofs rather than generalities and platitudes. The fact that purchasing a new $10,000 printer will allow us to print more elegant documents may sound nice, but what are the bottom-line numbers as to whether we will recapture our initial investment?

Accept Criticism Graciously

Some managers are good at giving negative feedback and some never learn how. Even poorly delivered negative feedback, however, provides an opportunity to grow and learn. New employees in an organization learn its rules, methodologies, and its culture through getting negative feedback on their efforts, often related to things they have written (Katz 1999). For a technical communication manager who takes pride in his/her communication skills, to receive negative feedback about a document from a non-technical communicator manager can be very trying. Nonetheless, communication managers should assume that their manager is using the feedback as a way of teaching them and increasing their skills, no matter how well or how badly the feedback was delivered. Further, the feedback will help the technical communication manager learn about the organization's methods and culture.

So long as feedback from one's boss is positive in its intent even if not smoothly delivered, a technical communication manager should pay attention to it and learn from it. When it seems to come from other motives, however, the technical communication manager must try to figure out what those motives are. Does the boss simply dislike technological communication and wish that the function was located elsewhere? Does the boss specifically dislike the particular communication manager. In this case the technical communication manager should probably start looking for another job or for a way to move the technical communication function to another manager's department.

If All Else Fails, Leave

Working for a boss who is universally acknowledged to be a loser has a seriously negative impact on a technical communication group. People often identify those who work under a bad manager as being equally poor at doing their jobs. Not only do they lose credibility as individual contributors, but their functions also lose credibility. Given the difficulty in many scientific and technological environments of gaining credibility for technical communicators in the first place, having to also fight through a second-rate manager makes that battle all the more difficult.

Also, working for a lousy manager takes a very heavy toll. If you are the only technical communication manager and you report to someone who is not liked or trusted or given any credibility, your working life is going to be much, much more difficult. Even if the boss is viewed positively throughout the rest of the organization but is someone who is unsympathetic to the goals of technical communicators, your day-to-day existence at work is going to be stressful.

Employees should be careful about falling into the "it will get better" syndrome, not dissimilar to what victims of abusive spouses often experience. Because one has made a choice, one can become determined to make it work. The psychologists call this phenomenon "cognitive dissonance," wherein people who have made a decision cling to it harder and harder as it becomes more and more obvious that the decision was a poor one.

The stress of working for a manager with whom you do not get along well or who has little credibility takes both an emotional and physical toll on a technical communication manager. What is already a tough job can become an impossible one. If you have tried everything else and you cannot alter the conditions under which you and your manager work together, consider going somewhere else. Perhaps there is another position within the organization. If not, look externally. It is simply not worth it to risk your career and your health because you are working for a lousy manager.

When You Find a Good One, Hang On

A good boss makes a major difference in the way you work and in how you feel about work. When you find a good one, take advantage of all of the coaching and mentoring that you can. Even if the boss in from a non-technical communication field, you can still learn much concerning other areas of management.

Unless your organization does only technical communication work, there will be some level of technical communication management that reports to someone who did not come from the communication arena. Having a non-technical communicator boss who understands our mission and our goals and who sympathizes with them is a blessing. You should still assume that you are going to have to educate that person about how we work and what our goals are, but it certainly makes the job of doing so easier.

Above all, when you have a manager who champions your discipline, demonstrate your gratitude in explicit ways. This does not mean toadying or "brown-nosing" but rather open, frank discussions about how important it is to you and to your people to know that the boss supports the kind of work they do and the contributions they make.

References

Bryan, J. G. 1994. Culture and anarchy: What publications managers should know about us and them. *Publications Management: Essays for Professional Communicators*. Edited by O. J. Allen and L. H. Deming, 55–67. Amityvillle, NY: Baywood.

Buckingham, M., as interviewed by Polly LeBarre 2001. Marcus Buckingham thinks your boss has an attitude problem. *Fast Company*: 88–98.

Covey, S. R. 1990. *Principle-centered leadership*. New York: Fireside.

Dobson, M. S. and D. S. Dobson. 1999. *Managing up: 59 ways to build a career-advancing relationship with your boss*. New York: Amacom.

Gabarro, J. J. and J. P. Kotter. 1993. Managing your boss. *Harvard Business Review* 71(3): 150–157.

Hackos, J. T. 1994. *Managing your documentation projects*. New York: John Wiley & Sons, Inc.

Katz, S. 1999. *The dynamics of writing review: Opportunities for growth and change in the workplace*. Norwood, NJ: Ablex.

Plung, D. L. 1994. Comprehending and aligning professionals and publications organizations. *Publications Management: Essays for Professional Communicators*. Edited by O. J. Allen and L. H. Deming, 41–54. Amityville, NY: Baywood.

Taylor, N. 1989. Managing management. *STC 36th Annual Conference Proceedings*. Society for Technical Communication.

For Further Information

Dobson and Dobson's book, *Managing Up: 59 Ways to Build a Career-Advancing Relationship with Your Boss* (1999), from the American Management Association, provides solid conceptual and practical advice.

A search on the Web under 'managing your boss' will yield dozens of articles. Two of the better ones are "Effectively Managing Your Boss" by Paula Moreia at http://www.certmag.com/issues/nov02/dept_techcareers.cfm and "Managing Your Boss," a *Harvard Business Review* classic by John J. Gabarro and John P. Kotter at http://www.copeland-mcdonnell.co.uk/managingboss.htm. The latter article also contains a section on the importance of understanding and managing yourself before you can successfully manage a relationship with your boss.

The American Management Association offers an on-line course led by Peter F. Drucker entitled *Managing Your Boss*, at http://ama.corpedia.com/welcome/product.asp?product=8103.

Questions for Discussion

1. How well do you get along with authority figures? What assumptions do you have about people in management?
2. What do you think it is like to work for someone like your boss's boss? What constraints does your boss work under? What priorities and results is he/she held responsible for? How is his/her performance measured?
3. Have you ever engaged in activities that you knew would cause you to lose credibility with a manager? Why did you do them?

4. How do you communicate problems to your boss? Is it his/her responsibility to fix them, or yours? Do you believe that by bringing up the problem you have done your job?

5. What are some of the rhetorical methods you can consider when delivering news that your recipient does not want to hear?

6. Think of an example of a time when you have successfully delivered bad news to your boss. Think of an example of a time when you have not been successful.

7. Have you ever procrastinated or even refrained from delivering bad news, simply because you did not want to confront the problems it would cause?

8. Is problem solving management's job or everyone's? Why should you go out of your way to help managers solve problems when they are getting the high salaries for doing so?

9. Why do managers sometimes seem to allow problems to go unfixed? Is it because they are clueless that the problem exists? Is it because they simply do not want to be bothered? Because they do not have the resources to do anything about it? Could there be other reasons, such as inter-group politics or some preference on the part of their boss?

CASE 7

Problem Solving and Communication

The Management Situation

As the new technical communication manager at Aardvark, you do not want to make too many waves in your first couple of weeks. However, you have learned that the network printer available to the technical communication group is the same one that is used by the company's public relations group. That group is extremely busy, with many deadlines. Every time they need to meet deadlines, they require that your group not print anything, even if it means that you will miss deadlines. They have also arranged with the network administrator for their print jobs to take priority in the queue over everyone else's. The people who have been doing Aardvark's documentation express extreme frustration over this, stating that they lose 5–10 hours per week of productive time while they wait to be able to access printers. Further, they say that waiting for the PR group has caused them to miss on-time delivery of review drafts and of final documents.

They propose several possible solutions, including (a) buying an inkjet for each technical communicator, (b) buying a high-speed, color, laser printer for the technical communicators, (c) having a showdown with PR management and demanding equal printing time and access, and (d) seeking upper management intervention to solve the problem.

You realize that you are going to have to recommend a solution quickly to your boss because you plan to hire several more communicators in the coming weeks and months, thereby putting even more strain on the printing resources. You also have some major deadlines coming, which means that you simply cannot allow printing concerns to prevent you from meeting them.

The budget for buying hardware is very tight, so you are going to have to prepare a compelling defense for the solution you arrive at for the printer problem.

The Assignment

Prepare a formal proposal to your boss about the printer problem and your proposed solutions for it. Read Chapters 4 and 7, and follow the ideas there for structuring your proposal.

Your boss likes to see numbers, so you are going to need to show what it will cost to do nothing and what each possible solution will cost. You should also indicate which solution you favor, and you should include a plan for implementing that solution. While the memo is to your boss only, you will also need to consider the other people involved in the printer issue.

Helpful Hints

Equitrac includes a printing cost calculator for computing cost per page and annual operating cost at http://www.metrics.com/Products/PAS/calculator.html.

Get current printer prices and reviews at http://www.zdnet.com and http://www.pcworld.com.

Evaluation Criteria for Case 7

Does your proposal follow a structure that leads to a compelling argument for the action you favor? Does it show the costs associated with doing nothing and with each possible solution?

Is your proposal appropriately aimed at your manager, who is a "big picture" person who does not like to deal with details, but who also likes to see numbers?

Does your proposal effectively lay out a convincing implementation plan and budget for solving the problem?

Does your proposal ask for permission to proceed?

CHAPTER 8

Other Technical Communication Management Responsibilities

Translation/Localization

With global trade expanding at a rapid rate, many technical communication managers are faced with getting their documents translated and localized for multiple countries. Merely translating a document literally might not work, as the methods and approaches used to prepare the documents for a United States audience may not be effective in some other cultures. Further, a literal translation may come out stilted and ineffective in some languages. Hence, it is necessary not only to translate but also to localize, which means that the contents and graphics must be altered as necessary so that they are effective with the desired audiences. This can be a major undertaking requiring considerable time and expense unless we take steps to accommodate it at the outset of a project.

Planning for Translation/Localization

Just as technical communicators prefer to be involved in a project from its onset, so do translators. Granted, at the beginning of the project there is nothing for them to translate. However, they can begin to become familiar with the product or service, with some of the terminology used to describe it, and with the basics of how it will function. Sending translators design specifications, documentation plans, and any other planning materials related to the project will help them prepare cost and time estimates and will help them prepare for the translation effort that is coming. Sending them drafts also helps them to plan their approach, even if you must ask them not to begin translating because you fear that the drafts will be revised considerably.

Be aware when estimating and planning for translation that page counts can change considerably when English is translated into another language. Because English is a relatively compact language and one where we often make up new words, translated texts can often be 10–50 percent longer than the equivalent English text. You may have to make allowances for this when planning document sizes to go in shipping cartons or mailing envelopes. You also want to be aware of it when providing page estimates to translators. One firm made "page-true" translations by increasing the size of the English version font and by adding more leading and white space to the English version (Walmer 1999). Hence, each page in each language's version had the same information, which could be helpful for customer service desks and for future revision efforts.

Good

It can take weeks or months to translate a technical document. If you need to test it with its intended audience, the time span becomes even longer. Because you cannot begin final translation until your text is more or less final, your translated versions are normally not going to be ready when the native version is. If the translated versions must ship at the same time as the native version, you will have to get the translators involved very early in the process and perhaps consider having them work on-site with the communication team.

Translation should be regarded as a sub-project, with milestone dates and deliverables carefully defined (Spalink 2000). If you are going to test the translated version, it is a good idea to ask for a segment of the translation as soon as possible, perhaps the first section or a key section dealing with a difficult set of tasks. This will give you the opportunity to begin testing and to establish "proof of concept" that the translator is meeting the needs of the audience.

If documents are initially written with subsequent translation in mind, the time and cost for the translation can be greatly reduced. Many communicators are trying to use what is called Simplified English or Simplified Language (Sanderlin 1988; Hoft 1995; Fernandes 1995), wherein the document is written using as small a vocabulary as possible and various rules (Weiss 1998) are followed to make the structure of the sentences and paragraphs simple and easy to translate (subject-verb-object structure, no complex sentences, etc.). In some cases, organizations have been able to pare their list of words used in documents down to 200–300 (Peterson 1990). While various authors champion various methods, the primary ones that are mentioned repeatedly include:

- Use the simplest, shortest words that will work
- Use simple sentence structure with active voice verbs
- Do not use synonyms; use the same word every time
- Avoid metaphors, similes, figurative language, idioms, and slang
- Prepare a glossary for the translator that defines terms

good

Depending on the language and culture of an intended audience, other methods may be needed, including taking care that graphics are amended so that they are effective (Horton 1993; Forslund 1996; Hager 2000).

For more information on translation and localization, see Edmund Weiss's (1998) excellent list of tactics and Hoft's *International Technical Communication: How to Export Information About High Technology* (1995).

Hiring a Translator

Many translation firms advertise in STC's *Technical Communication* and *Intercom*. To hire a translator, you should prepare a request for proposals (RFP) that is as detailed as possible. You can usually adapt your documentation plan so that it works as an RFP, adding details as necessary pertaining to contractual arrangements, additional delivery dates and milestones for the translation, and additional assumptions and expectations regarding the translation. The translator will want to know the size of the document, usually in the number of words rather than in pages, which can vary considerably. If you have an example of an earlier project or a competitor's product that is about the same size and scope as your current project, you can include it with the RFP to give the translators a better idea as to how to estimate. The earlier you can get the translator involved the better, even if that involvement is only a few hours a week at first.

When you get proposals back from the translation firms, you want to look carefully at the resumes of the proposed translators. Most of us who have used translators would agree with Ulijn (1996) that they should be native writers in the client language, should be natives of the client culture, and should have a thorough knowledge of the subject matter. You are trying to find a translator who fits those criteria as closely as possible. If you have to pay extra to meet those criteria, it is a good idea to do so. Taking a chance on an unqualified translator could wind up costing you much more than you initially save because of lower translation costs. You also want to look carefully at the requirements and assumptions of the translator to make sure that you can meet all of them. If they require a glossary, do you have the time and resources to prepare one? If they require two copies of each graphic, one with callouts in your native language and one that has no callouts, do you have someone who can prepare them?

Once you have hired the translation firm, you need to include the translation dates in your project schedule, preferably not only in the technical communication schedule but in the overall project schedule, so that everyone stays informed as to the translation's progress and so that you are reminded when you need to send information to the translator.

How do you check the translations you get back to see if they are accurate? Obviously, the best way is to have a native speaker read the documents and try to use them to accomplish the tasks that your audience will be trying to accomplish. This can be done informally or it can be set up with formal usability testing methods. Without some kind of testing, you risk sending out a translation that could be ineffective, misleading, or even dangerous. Further, you risk the extra cost and effort of having the document re-translated if it is ineffective for its intended audience. While we might get away with not testing documents intended for our own native audiences, we will rarely be able to do so with international audiences.

Preparing Proposals and RFPs

Proposals

Technical communication managers usually have to prepare both proposals and requests for proposals. Your proposals might take the form of project estimates or documentation plans, but they are still basically proposed levels of work at a proposed price for your efforts on a project. In fact, it is a good idea to consider any request for your services as an RFP, even if the requestor casually asks you what it would cost to get a 100-page document done. If you casually answer such a question with an amount, that amount becomes your price even if later calculations show that the actual cost will be much higher. If the requestor also casually asks by what date you can have the document completed and you casually answer June 1, that becomes your deadline. Therefore, you should always answer such questions with a request for more information regarding the scope, purpose, development schedule, and any specs or requirements that are available for the project. You cannot possibly give an informed estimate and deadline without such information.

You also want to respond to questions and requests with a document that not only provides the information requested, usually a cost and a date, but that also describes the assumptions you have made to arrive at the costs and dates you are committing to. Scientists and engineers often grossly underestimate the time it takes to prepare quality documentation. Therefore, you want to explain as fully as possible how you arrived at your numbers, what your assumptions are about the nature and scope if the project, and the requirements that you have for access to information and to subject matter experts if you are to meet the budget and schedule figures you are quoting.

A complete technical communication proposal contains the sections that Hackos (1994) suggests for a project plan, plus some additional sections:

1. Executive Summary
2. Benefits to client of using your services
3. Objectives of the overall project and of the documentation
4. Audience/task analyses and design implications
5. Documents/services to be delivered
6. Estimate of the project's scope and complexity
7. Estimate of time and budget required
8. Estimate of the resources required
9. Schedule of milestones (including client responsibilities)
10. Assumptions
11. Roles and responsibilities of TC and client team members (including reviewers)
12. Production plan

13. Translation and localization plan
14. Usability and validation testing plan
15. Maintenance plan

While you might not include all of these sections in every proposal, especially those for internal groups, you should cover as much of this information as you possibly can. Why go to this much trouble? Again, many people do not understand the cost of developing good documentation (much less of developing online information and Web sites). They believe that if they can write a page of rudimentary directions off the top of their heads in 15 minutes that you should be able to write an entire document at the same pace. And they think that because their 14-year-old nephew developed a Web site overnight that you should be able to do so, too. A proposal helps to educate them as to how technical communicators add quality to the document preparation process by conducting extra activities such as audience analysis, task analysis, substantive editing, usability testing, and more. The proposal also serves as a means for avoiding misunderstandings in the future, when the scope of the project changes. And it further serves as a form of protection so you cannot be accused in the future of running up the development costs when they turn out to be higher than the client anticipated. In short, a technical communication manager should never commit to work, even internally, without a proposal that spells out the costs, the dates, and the assumptions behind them.

It is outside the scope of this book to include complete instructions for writing proposals. For more information, see the Association of Proposal Management Professionals' Web site at http://www.apmp.org/home.html. Also, Peterson (1998) provides some helpful tips on preparing technical communication proposals.

Requests for Proposals (RFPs)

Many technical communication managers must occasionally hire contract writers to fill in when a project requires extra resources but does not warrant hiring a full-time employee. Communication managers might also contract for services from graphic artists, multimedia developers, programmers, production houses, printers, and other vendors. In those cases, it will be necessary to prepare a request for proposals, so that you can get competitive bids from more than one vendor. Even if you decide to "sole source" a project (using only one vendor without competitive bidding), it is still a good idea to write an RFP that spells out your needs and your expectations for the project. The RFP and the proposal(s) you receive provide the basis for a contract to do the work. In fact, on smaller contracting jobs the proposal often serves as a contract, with both sides signing to verify that they agree to the terms and assumptions spelled out in the proposal.

RFPs can range in formality from one page to hundreds of pages (especially in the case of large government contracts). For most technical communication managers, the RFP can be one to three pages. Durham (1999) says that an RFP should have the following sections:

1. Cover letter
2. Introduction

3. Background
4. Focus
5. Selection criteria
6. Time line and details
7. Method
8. Details
9. Evaluation
10. Budget
11. Proposed outcomes/deliverables
12. Legal and ethical issues
13. References and supporting material

To get the services of a temporary contract writer on an hourly basis, you would probably need only a brief introduction, a sentence or two of background and focus, a list of the criteria you will use, the end date of the contract period, a few details about the project, how much you are willing to pay (usually as a range), and the deliverables that the contractor will work on (although for an hourly contract deliverables are not necessary). For a larger project or for one where a specific deliverable will be made to you, it becomes more important to include each of the sections and to spell out in detail what is expected for the deliverable. The greater the detail you can provide, the less opportunity there is for misunderstandings later.

Before writing an RFP, check with your organization's purchasing department to see if they require any special guidelines and/or standard forms . You may have to do more paperwork than you want to, but purchasing departments often require such documentation to ensure that purchasing decisions are made fairly, ethically, and impartially. If you want to use a specific vendor, called a sole source, you may have to write a justification explaining why that vendor is uniquely qualified to provide the services.

Most technical communication contract work is done on a temporary basis. Each time we need a certain service, we must write another RFP, wait for proposals, evaluate the proposals, and select a vendor. For some ongoing services, it might be preferable to use what is called a retainer. Here, we retain the services of a firm, usually for a specified period of time with a guaranteed minimum amount of work. For example, if we know that we will need about a quarter of a person's worth of Web development work over the next year, but at sporadic intervals, we might retain one firm to provide a guaranteed 500 hours of Web development services through December 31 (or whatever end date is appropriate). The firm would guarantee that within 24 or 48 hours advance notice they could supply a Web developer to work with us for several days at a time. Even better, we can schedule regular intervals of work, such as monthly maintenance on our Web site.

Vendor Relationships

Having a group of vendors with whom you have worked in the past and found to be reliable can be a valuable asset for a technical communication manager. As our work becomes more electronic and multimedia-based and less paper-based, we will need to assemble teams with widely divergent skills. For some of the skill sets that are used only occasionally, it does not make sense to hire a full-time employee. So, we are likely to need contractors more often in order to assemble the teams our projects will demand. Having good working relationships with vendors who can supply the people and services we need is simply good business practice.

However, there is a fine ethical line between having those good vendor relationships and excluding other qualified vendors from bidding on projects. Purchasing departments often have policies that guard against employees becoming too close with one or two vendors. One obvious reason is concern about "kickbacks," wherein an employee receives compensation from the vendor in exchange for giving the vendor ongoing business. Some technical communication managers, especially at larger firms, may have printing contracts that run into hundreds of thousands and even millions of dollars. While most of us would never entertain the idea of taking cash kickbacks, we might accept other gifts that have the appearance of impropriety. For example, if your printer vendor gives you tickets to a ballgame or the opera (worth perhaps $50–$100), is that a kickback? If they send you a smoked turkey at Christmas (worth, say, $50), is that a kickback? Some organizations forbid acceptance of any item, no matter how trivial, from a vendor, to prevent even the appearance of impropriety. While most of us cannot be bought off for a pair of tickets or a turkey, we must still guard against becoming too cozy with individual vendors.

One of the best ways to maintain good relationships with vendors is to ensure that they get paid promptly. When you get an invoice from a vendor, promptly approve it and send it to your accounts payable group. If necessary, follow up to make sure that the vendor gets paid within a reasonable period. I once had the embarrassment of a vendor calling after six months inquiring why they still had not been paid. Phone calls to the accounts payable group did not achieve a resolution. I finally had to go to the accounting department, where we discovered that the account had been handled by an employee who had left, and where we found the invoice at the bottom of a stack of papers on his desk.

Another good idea is to visit the facilities of vendors whom you hire, especially if there is production or printing work involved. You will learn more about their type of work from getting the "guided tour," and you will also begin to see which organizations have high-quality operations and which do not. Further, when last-minute crises arise, you will have a better idea about who might be able to help you most effectively. Also, a vendor who has met you fact-to-face and has spent some time with you is more likely to respond when you have special requests or need extra help.

Marketing/Sales

Whether we like it or not, all of us are involved in sales. We must constantly sell our services to internal groups and, in some cases, to external groups. Even in situations where internal groups are "captive" clients who are required to use our services, we should still adopt a sales-based mentality in our relationships with those clients. We should sell them on the fact that they are receiving a quality product or service at a reasonable price and that we are the best possible means for them to get those products or services. Because there are so many firms offering technical communication services, internal management people are going to wonder constantly if they wouldn't be better off outsourcing the work. The best way to prevent that from happening is to remind them constantly that they are getting a better product than they could from the outside, even if the cost appears to be higher. Further, we should also work to persuade them that what we do is inextricably joined with the very nature of the services and products that the overall organization produces.

Competitive Analysis

It is a good idea for any technical communication group to do a competitive analysis. What alternative solutions could your organization look to for technical communication work ? Make a list of all you can think of. It should normally include each of the following:

- Do not do documentation at all
- Have scientists and engineers write the documentation
- Have secretaries or administrative assistants "wordsmith" the specs and use those as our documents
- Outsource all documentation
- Hire technical communication project managers, but outsource all of the development work
- Have a dedicated technical communication group

Your competitors, then, are internal personnel at various levels and external technical communication firms. For your competitive analysis, list the advantages and disadvantages of each solution, including the costs. Notice that to upper management, the first three solutions have essentially no cost, as the time that engineers or secretaries spend on documents will essentially be hidden. Notice also that having your technical communication group will almost always look like the most expensive solution. To managers who are looking for every possible way to pare costs, this makes technical communication a conspicuous area for consideration. If we look like a "cost center," we are likely to have to fight to avoid having someone select one of the apparently cheaper solutions. So far, our competitive analysis has shown that we are the most expensive solution. It had better also show that we are the highest quality solution. How do we sell the idea that we can provide higher quality than the other alternatives?

Selling Technical Communication Services

One of the best ways to stress the quality of an internal technical communication group is to resort to the value-added analyses described in Chapter 3. However, there may be a danger in simply relying on dollar amounts to prove value. As the formula used in Chapter 3 shows, intangibles, which are not easily quantified, add considerably to the extra quality that an internal group contributes. The following arguments help describe ways to sell technical communication services, both internally and externally.

Stress the Unique Rhetorical Processes Used by Technical Communicators

Schriver (1997) discusses three main conceptual approaches to documentation design: the craft, the romantic, and the rhetorical. Many managers and potential clients have the craft concept of technical communicators, that we "wordsmith" documents and "pretty them up," but that we do not add much value otherwise. As Schriver says,

> The overly simple views of writing that were fostered by the teaching of the craft tradition and through the regimentation of business practices during the first half of the century have had a lasting detrimental effect on the professional development of writers in organizations. Because these reductive views are still widely held, writing in some companies is construed as a marginal activity that adds to an organization's costs but that contributes neither to "the bottom line" nor to the quality of products and services (p. 77).

This craft or "wordsmithing" concept should be countered every time it is uttered. Even better, it should be averted right from the start with a detailed explanation of the processes we follow to create high quality documents. Anyone (including engineers and secretaries) can "pretty up" a document, and those who made A's in English believe that they are as good at it as professional communicators, especially if those communicators are perceived to be working in the craft tradition. Not everyone, however, knows how to perform audience and task analyses, how to link those analyses to the organization's products and services, and how to design appropriate documentation libraries and structures to meet the requirements of each rhetorical situation. When technical communication managers describe such processes in detail, they help to establish themselves as professionals with unique skill sets that not everyone possesses. They establish that technical communication is more than "wordsmithing" and that it requires specialized knowledge and skills based on education and on experience.

That set of specialized skills will not be evident if documents are developed by engineers or secretaries. Nor will it be evident when using external contractors who have had little or no contact with the audience or with the products and services being developed for that audience. The strongest argument that technical communicators can make about their value and their professionalism is that they have a unique set of standard processes they use to ensure high quality, usable, customer-oriented products and services.

Emphasize Customer Contacts/Advocacy

Another strong argument for technical communication services is that they provide the last line of defense for the customer and that technical communicators, in fact, see themselves as customer advocates who prevent engineers and programmers from calling an on/off switch a "process implementation device." We should emphasize that this goes beyond mere "wordsmithing" and editing. It involves applying human-factors principles to the way people interact with machines, computer software, Web sites, and other processes, and it involves knowing how to label and conceptualize such interactions to work most effectively with the customers who will use them. This argument often receives support from people in sales/marketing and in customer support/relations groups, who often see more readily than others the value that communicators can add.

Emphasize That Documents Are Part of the Product

Another powerful argument that helps sell technical communication work is that the artifacts communicators create are not "support" but are an integral part of a product. In some cases this is patently true, as in the example of a Web site, where the site itself may be the product. In the case of a software product with on-line help built in, many programmers freely admit that the documentation becomes an integral part of the product. But even in the case of a piece of hardware with a separate paper document sent along with it, we can still contend that the overall product experience of the customer involves using both the hardware and the document, and that they would be unable to use the hardware if they did not have a document. In essence, the product then is worthless without the document.

With software, a strong case can be made that the user interface (the screens) and the documents (whether on-line or on paper) *are* the product, from the customer's point of view. The customer initially installs the software from a disk, tape, CD, or a download. The CD is put away and never seen again, unless something goes wrong. From that point on, the customer's only interaction with the product is with the screens and the documents. From the customer's point of view, then, those things *are* the product, and the code itself is invisible. From this perspective, technical communicators and human factors experts have created the product, and the programmers have provided a support service. Next time you see one, say thanks.

This can be a very powerful argument in a software or web development firm, where technical communication is often assumed to be a support service. While some programmers will not like the argument, management, along with people in marketing and customer support, will often agree with the idea and will ensure that technical communicators receive enough funding to make the product as successful as possible.

Describe the Efficiencies of Working Internally

While using outsourced technical communicators might seem less expensive at first blush, there are cost considerations that make it more expensive. If the outsourced communicators are off-site, they lose the advantages in time and efficiency that internal communicators naturally have. While an internal communicator knows the right people to ask questions of and can walk down the hall and ask them, the external contractor might have to send an e-mail and wait hours or days for an answer. Internal communicators are available instantly for ad hoc meetings, status updates, quick emergency fixes, etc., whereas outside contractors must be called or e-mailed and invited in for an official meeting, all of which takes more time on the part of the internal project managers and other development staff.

Internal communicators quickly learn the organization culture, guidelines, and requirements for how work gets done and how to get authorizations, approvals, etc., whereas external communicators may have to rely on project managers and other internal people to get such matters handled. Further, outside contractors will not have the same commitment to the organization and to seeing that it produces documents of the highest possible quality. It is no secret to most technical communication managers that our employees work much overtime, usually for free. Outside contractors will want to be paid for overtime, which in itself may be enough to overcome the apparent savings generated by external sources.

By the same token, if you are trying to sell technical communication services to an external organization, request that the communicators work at their site if at all possible. At the very least, try to get a lead communicator or project manager on the client's site, so that you gain some of the advantages of having someone co-located with the scientists and engineers with whom the work will be done.

Emphasize Greater Accuracy

An internal technical communication group is more likely than an outside contractor to become familiar with the product or service line, with the conceptual underpinnings about how the products work, and with the most appropriate methods for communicating with customers. Many technical communicators eventually become subject matter experts about the products and services they cover, often contributing much more to project teams than simply developing the documentation. While these contributions may not show up on a spreadsheet, their absence will become apparent when they are gone. Ask scientists or engineers if they want to train a new communicator as to how their various technologies work.

Internal technical communicators will usually have much greater access to subject matter experts than will external contractors, who will have to use phone calls, e-mail, and site visits to get information from the SMEs. The internal communicators will often have learned the most effective methods and approaches for working with various SMEs, and they will be considerably more efficient (and use less SME time) than will external contractors. External contractors are more likely to give up on getting a piece of information and to put a "best guess" in a document, assuming that it will be caught during review if something is wrong. An internal communicator, who can simply walk down the hall and ask a quick question, will be more likely to follow up on ambiguous information and to achieve more accurate results.

Link Technical Communication to the Strategies of the Information Age

Another persuasive argument about the importance of technical communication has to do with the very nature of the information age. While it is no doubt true that many of the proven business principles still hold, the information age means that one of the things that distinguishes one organization from others is how well that organization conceptualizes, creates, designs, develops, and delivers information to its customers. An organization that believes information is not a part of its core business and that it makes good business sense to outsource experts in information development risks destroying any competitive edge that it has and, indeed, risks its very existence.

This is more obviously true with Web-based organizations and with those that provide all of their information on-line. However, it is also true of organizations that provide documentation for hardware and for services that are not electronic. Customers today expect the same kind of instant information for these products that they can get on the Web for software products. They are simply not going to be satisfied with inefficient, inaccurate, and untimely information. Every organization needs to address these customer preferences. To do so, it is critical to have the best possible people using the best possible methods for communicating with customers. Eliminating the professional communicators or trying to accomplish the job using amateurs will cost more than the apparent short-term savings.

A corollary is that customers will judge the ethos (the reputation and authority) of an organization by how well it communicates and by how professional its methods for doing so appear. An organization that uses amateurs or outsiders to create its communications risks losing credibility for its products, its services, and its brand.

References

Durham, M. 1999. "Finding the perfect match—writing requests for proposals." *STC 46th Annual Conference Proceedings*. New Orleans: Society for Technical Communication.

Fernandes, T. 1995. *Technical communication in the global community*. Boston: Academic Press.

Forslund, C. J. 1996. Analyzing pictorial messages across cultures. *International Dimensions of Technical Communication*. Edited by D. Andrews, 45–48. Arlington, VA: Society for Technical Communication.

Hackos, J. T. 1994. *Managing your documentation projects*. New York: John Wiley & Sons, Inc.

Hager, P. J. 2000. Global graphics: effectively managing visual rhetoric. *Managing Global Communication in Science and Technology*. Edited by P. J. Hager and H. J. Scheiber, 21–43. New York: John Wiley & Sons, Inc.

Hager, P. J. and H. J. Schriber. 2000. *Managing global communication in science and technology*. New York: John Wiley & Sons, Inc.

Hoft, N. L. 1995. *International technical communication: How to export information about high technology*. New York: John Wiley & Sons, Inc.

Horton, W. 1993. The almost universal language: graphics for international documents. *Technical Communication* 40(4): 682–693.

Peterson, D. A. T. 1990. Developing a simplified English vocabulary. *Technical Communication* 37(2): 130–133.

Peterson, T. A. 1998. Tips for winning proposals. *Intercom* 45(8): 7–9.

Sanderlin, S. 1988. Preparing instruction manuals for non-English readers. *Technical Communication* 35(2): 96–100.

Schriver, K. A. 1997. *Dynamics in document design.* New York: John Wiley & Sons, Inc.

Spalink, K. 2000. Improving cost-effectiveness in the documentation development process through integrated translation. *Managing Global Communication in Science and Technology.* Edited by P. J. Hager and H. J. Scheiber, 179–202. New York: John Wiley & Sons, Inc.

Ulijn, J. M. 1996. Translating the culture of technical documents: Some experimental evidence. *International Dimensions of Technical Communication.* Edited by D. Andrews, 69–86. Arlington, VA: Society for Technical Communication.

Walmer, D. 1999. One company's efforts to improve translation and localization. *Technical Communication* 46(2): 230–237.

Weiss, E. 1998. Twenty-five tactics to internationalize your English. *Intercom*(45): 11–15.

For Further Information

The first two sources to consult for translation and localization are Nancy Hoft's book, *International Technical Communication: How to Export Information About High Technology* (1995) and Peter J. Hager and H. J. Scheiber's book, *Managing Global Communication in Science and Technology* (2000). JoAnn Hackos, in *Managing Your Documentation Projects* (1994) and Karen Spalink, in her chapter, "Improving Cost-Effectiveness in the Documentation Development Process Through Integrated Translation" (2000), both stress the importance of treating translation and localization as a sub-project, with milestone dates and deliverables well defined and project management techniques closely followed.

Reducing translation costs through using Simplified English or Simplified Language has been treated by numerous authors, including Nancy Hoft in her 1995 book, Stacey Sanderlin in her article, "Preparing Instruction Manuals for Non-English Readers" (1988), and Tony Fernandes in his book, *Global Interface Design: A Guide to Designing International User Interfaces* (1995). In "Twenty-five Tactics to Internationalize Your English" (1998), Ed Weiss offers a list that has become a standard tool for writers concerned about translation.

William Horton, in "The Almost Universal Language: Graphics for International Documents" (1993), Charlene J. Forslund, in "Analyzing Pictorial Messages Across Cultures" (1996), and Peter J. Hager, in "Global Graphics: Effectively Managing Visual Rhetoric" (2000) all discuss the importance of amending graphics as well as text so that the images function properly in the new language and culture.

For writing proposals related to technical communication projects, JoAnn Hackos's *Managing Your Documentation Projects* (1994) includes detailed outlines for information and project plans, which provide most of the parts of a traditional proposal. Tracy A. Peterson provides concise ideas for preparing good technical communication proposals in "Tips for Winning Proposals" (1998). The Association of Proposal Management Professionals has a helpful Web site at http://www.apmp.org/home.html. Marsha Durham provides valuable assistance for writing requests for proposals in "Finding the Perfect Match—Writing Requests for Proposals" (1999).

Questions for Discussion

1. Look through the documents you have at home for appliances, electronics equipment, camera equipment, computers and accessories, etc. See if you can find an example of a poorly translated manual. What effect does the translation have on the usability of the document? What effect does it have on your perception of the quality of the product and of the company selling it?

2. Hasn't English become a second language for nearly everyone? Can't we assume that our audience, especially for technical and scientific products and services, will know English well enough so that translation is no longer necessary?

3. If you have spent considerable time and money getting your document translated into Swedish, and you have no Swedish speakers on your staff or in your town, how do you test the document to see if it will work with its intended audience?

4. What are the rhetorical implications for a technical communication proposal? What are readers of such a document looking for? How do you write a proposal when you know that your price will not be the lowest one? When you know that it will be the highest one?

5. Why bother to write a proposal or a documentation plan for an internal project where the internal clients are forced to use your services?

6. Why go to the time and expense of writing a request for proposal, reading through the resulting proposals, choosing a vendor, negotiating the final terms, and signing a contract when all I want to do is hire a couple of contractors for three months? Why can't I just call my friend Melissa down the street and give her the contract, which will cut out the bureaucracy of the purchasing department and will expedite things by three weeks?

7. Why should a technical communication group have to sell its services within its own organization? The scientists and engineers do not have to do so—why should we?

CASE 8

Writing a Proposal for Technical Communication Services

The Management Situation

As the technical communication manager at Aardvark Enterprises, you have been asked by the internal software development group to give them a quote for the cost to prepare on-line help and paper documents to support a new product they are about to begin developing. While all they asked for was a dollar amount, you realize that you should supply a full proposal that includes the assumptions behind your price estimate along with the other components of a technical communication proposal or documentation plan.

Aardvark is an energetic, start-up company that puts out a line of software products aimed at several niche, vertical markets in the financial field. Because the company is so young, it does not have any established project management methods for its technical communication efforts. Recognizing that they need to improve their documentation development process, management has promoted you into the technical communication management position.

Your first assignment is a new product development effort that is expected to require six months. The software product being developed will run under Windows and will allow users to track stock portfolios, buy and sell equities, obtain detailed reports about their holdings, and make decisions about future purchases. The product, called Midas, is aimed at the small investor rather than at institutions.

After assessing the initial requirements document for the new software product and conducting audience and task analyses with potential users, you have concluded that you will have to supply a help system with approximately 200 help topics, a getting-started document of about 50 pages, and a job aid in the form of a tri-fold brochure.

The project is slated to begin on January 1 and to take 6 months. The alpha version of the code will be ready on April 1, the beta version on May 15, and the final version on July 1.

You have been assured that, since this will be the first time that Aardvark has developed such a product, you will have ready access to the subject matter experts. Because this is your first new project at Aardvark, your team does not have experience working together. They do have experience using the software products you plan to use to develop the help and documentation. The product will be marketed only within your country at first, so no translation or localization will be required. You will not need to purchase any additional software or hardware for the project, but you will need to make two trips to do testing with potential users, costing $1,000 for each trip.

The Assignment

Write a proposal of 8–10 pages that contains each of the 15 sections listed in this chapter. You will first have to prepare estimates for each of the three documents, using the Technical Communication Estimation Worksheet from Chapter 3. Assume that your communicators are paid $30 per hour and that Aardvark's loadings bring the loaded rate for a communicator up to $100 per hour.

You will have to invent some of the details for the proposal, such as the names of communicators and of reviewers. You should include a schedule that shows when you will deliver milestone drafts and when you require reviewed drafts back.

Helpful Hints

Read Chapters 7 and 8 in Hackos's *Managing Your Documentation Projects*.

Read Peterson's "Tips for Winning Proposals" in *Intercom*, 1998, issue 8, pp. 7–9.

Read England's "Foregrounding the Customer in Technical Proposals" in the *Proceedings of the Society for Technical Communication 44th Annual Conference*, 1997, pp. 4–6, available at http://www.stc.org/proceedings/ConfProceed/1997/PDFs/0006.pdf.

For comprehensive information on writing proposals, see the Proposal Management Professionals Web site at http://www.apmp.org/home.html.

Evaluation Criteria for Case 8

Does the proposal make a persuasive case for using the services of your technical communication group?

Does the proposal clearly show the rationale for the types of documents you propose to develop?

Does the proposal include project costs and schedules?

Does the proposal make clear all of the assumptions on which the price and schedule commitments are based?

Does the proposal compellingly demonstrate to the client the benefits of using your services?

Resources for Technical Communication Managers

This chapter contains lists of information that should be helpful to technical communication managers. Obviously, training and translation companies come and go, as do Web sites and organizations. While the lists are as accurate as possible at the time of publication, you may have to do some searching to find Web site addresses later.

Books

This list contains only books dedicated to technical communication management. Obviously, many business and management books aimed at the general market would also benefit a technical communication manager, but it would be impossible to list the thousands of titles here. While many technical communication books contain chapters or sections about management issues, this list includes only technical communication books that are aimed exclusively at managers or that contain major portions dedicated to management questions.

Allen, O. Jane and Lynn H. Deming, eds. 1994. *Publications management for professional communicators*. Amityville, NY: Baywood.

Hackos, JoAnn. 1994. *Managing your documentation projects*. New York: John Wiley & Sons, Inc.

Magazines/Journals

Numerous journals serve the technical communication field, some devoted primarily to practitioners and some aimed almost exclusively at academicians. Table 9.1 lists some of the magazines and journals of interest to technical communicators. I have indicated for each journal whether it would be of more use to a practitioner or an academician, realizing that such judgments are often highly subjective. This table includes journals cited by Schriver (1994) plus others.

TABLE 9.1	Technical Communication Journals	
Journal Title	**Primary Audience**	**Contact Information**
Technical Communication	Practitioners	www.stc.org
Intercom	Practitioners	www.stc.org
Journal of Technical Writing and Communication	Academics, Practitioners	www.baywood.com
Issues in Writing	Academics	http://www.uwsp.edu/English/iw/
Journal of Computer Documentation	Practitioners, Academics	http://www.acm.org/sigdoc/
American Medical Writers Association Journal	Practitioners, Academics	http://www.amwa.org/
Human Factors	Academics, Practitioners	http://www.hfes.org/
Human-Computer Interaction	Academics, Practitioners	http://www.erlbaum.com/
IEEE Transactions on Professional Communication	Practitioners, Academics	http://www.ieeepcs.org/
International Journal of Human-Computer Studies	Academics, Practitioners	http://www.academicpress.com/ijhcs
Journal of Business and Technical Communication	Academics, Practitioners	http://www.sagepub.com/
SIGCHI: Journal of the Special Interest Group on Computer-Human Interaction	Academics, Practitioners	http://www.acm.org/sigchi/
Technical Communication Quarterly	Academics	http://www.attw.org/
Written Communication	Academics	http://www.sagepub/

TABLE 9.2	Technical Communication Training Organizations	
Training Org.	**Telephone**	**Web site**
Advacon	(919) 942-2322	www.advacon.com/content/index.cgi
Bright Path Solutions	(919) 656-2653	http://www.travelthepath.com/
Center for Information-Development Management	(303) 232-7586	www.infomanagementcenter.com/index.htm
Data Solutions Corp.	(800) 566-7900	http://www.datasol.net/
EEI Training	(888) 2LEARN2	www.eeicommunications.com
Langevin Learning Services	(800) 223-2209	www.langevin.com
Influent Tech. Group	(888) 333-9088	www.influent.com
Information Mapping	(800) 843-6627	www.infomap.com
NSight	(617) 354-2828	www.nsightworks.com
Padgett-Thompson	(800) 255-4141	www.ptseminars.com
Pubsnet Inc.	(508) 244-0272	www.pubsnet.com
Scriptorium	(919) 481-2701	www.scriptorium.com
SkillPath Seminars	(800) 873-7545	www.skillpath.com
Solutions	(800) 448-4230	www.sol-sems.com
User Edge	(908) 730-9164	www.useredge.com
Weisner Associates	(800) 646-9989	www.weisner.com
Winwriters	(800) 838-8999	www.winwriters.com

Training

Table 9.2 lists organizations and companies that provide training specifically in technical-communication-related areas.

Translators

For a comprehensive listing of translator societies, see http://www.translatortips.com/profsoc.html. For more information about translation and localization, see http://www.lisa.org.

Table 9.3 lists translation firms which have recently advertised in STC journals, plus a few others. Of course, there are many other translation firms and many fine translators who work as independent contractors.

TABLE 9.3	Translation/Localization Organizations
Translation/Localization Company	**Address**
Adams Translation Services	http://www.adamstrans.com
Bowne International	http://www.bowne.com/international/index.asp
Burg Translations	http://www.burgtranslations.com
Center for Technical Translation	http://www.cicenter.com
Harvard Translations	http://www.htrans.com
International Communication by Design, Inc.	http://www.icdtranslation.com
International Access/Ability Corp.	http://www.iac.com
International Translation Resources Ltd.	http://www.itr.co.uk
International Translation Solutions	http://www.intransol.com
JLS Language Corporation	http://www.jls.com
Lingo Systems	http://www.lingosys.com
Lionbridge	http://www.lionbridge.com
NCS Enterprises	http://www.ncs-pubs.com
Ralph McElroy Translation Company	http://www.mcelroytranslation.com
RWS Group translate.com	http://www.translate.com
SH3	http://www.sh3.com
Star-USA	http://www.star-usa.net
Trados Corp.	http://www.trados.com

Indexers

The American Society of Indexers provides valuable information related to indexing at http://www.asindexing.org/.

Organizations

The only organization devoted exclusively to technical communication managers is the Society for Technical Communication's Special Interest Group in Management, at http://www.stcsig.org/mgt/index.htm.

Table 9.4 provides a list of organizations to which technical communicators may belong, plus contact information.

TABLE 9.4	Technical Communicator Organizations
Organization	**Contact Information**
Society for Technical Communication	http://www.stc.org
ACM Special Interest Group in Documentation (SIGDOC)	http://www.acm.org/sigdoc
IEEE Professional Communication Society	http://www.ieeepcs.org
American Medical Writers Association	http://www.amwa.org/
Council of Science Editors	http://www.councilscienceeditors.org/
Board of Editors in the Life Sciences	http://www.bels.org/
American Society for Training and Development	http://www.astd.org/
Graphic Communications Association	http://www.gca.org/
International Association of Business Communicators	http://www.iabc.com/homepage.htm
International Society for Performance Improvement	http://www.ispi.org/
National Association of Government Communicators	http://www.nagc.com/
Usability Professionals Association	http://www.upassoc.org/
ACM Special Interest Group in Human-Computer Interaction	http://www.acm.org/sigchi/
National Association of Science Writers	http://www.nasw.org/
Localization Industry Standard Association	http://www.localization.org/
Public Relations Society of America Technology Section	http://www.tech.prsa.org/new/index.html
Association of Earth Science Editors	http://www.aese.org/
International Council for Technical Communication	http://www.intecom.org/
Association of Proposal Management Professionals	http://www.apmp.org/home.html

INDEX